# Digital Effects, Strategies, and Industry 5.0

This book discusses the increase in global competitiveness which challenges the manufacturing market to integrate design and product in order to improve quality and process. The book goes on to examine digital manufacturing technologies and critiques how they are transforming every link of the manufacturing value chain.

*Digital Effects, Strategies, and Industry 5.0* presents many different tooling processes that digital manufacturing utilizes such as artificial intelligence, automation and robotics, additive technology, human-machine interaction, and IoT. Digital manufacturing technologies and how they can transform every link of the manufacturing value chain, from research and development, supply chain, and factory operations to marketing, sales, and service, are examined within the book. Also included is coverage of Industry 5.0, the future, and how it is already starting a trend of change processes directed towards closer cooperation between man and machine, as well as systematic prevention of waste and wasting including industrial upcycling, along with case studies.

This book is aimed at professionals and students in the areas of manufacturing and processing, productivity improvement, environmental, engineering management, and information management.

# Digital Effects, Strategies, and Industry 5.0

Fabio De Felice
Antonella Petrillo

CRC Press
Taylor & Francis Group
Boca Raton  London  New York

CRC Press is an imprint of the
Taylor & Francis Group, an **informa** business

First edition published 2024
by CRC Press
2385 NW Executive Center Drive, Suite 320, Boca Raton FL 33431

and by CRC Press
4 Park Square, Milton Park, Abingdon, Oxon, OX14 4RN

*CRC Press is an imprint of Taylor & Francis Group, LLC*

ISBN: 978-1-032-29496-4 (hbk)
ISBN: 978-1-032-29497-1 (pbk)
ISBN: 978-1-003-30184-4 (ebk)

DOI: 10.1201/b22968

Typeset in Times
by MPS Limited, Dehradun

# Contents

# Preface

In the context of an ever-changing world, technological innovation and digitization have assumed a fundamental role in shaping our present and determining the future. This book, *Digital Effects, Strategies, and Industry 5.0*, aims to explore the vast landscape of innovative technologies that are transforming our lives, businesses, and society as a whole.

The opportunities offered by innovative technologies are immense. Digitization has made it possible to access an unimaginable amount of information and has broken down geographical barriers, opening up new horizons for communication and collaboration. Artificial intelligence, machine learning, and other emerging technologies are transforming the industry and creating new business models. Robotics and automation are revolutionizing manufacturing processes and improving operational efficiency. Blockchain is redefining trust and financial transactions. The Internet of Things connects everyday objects to the network, creating an interconnected and intelligent environment.

However, along with the opportunities, significant challenges also arise. Technological innovation advances at an accelerated pace, putting a strain on the ability of companies and institutions to adapt and take full advantage of the new possibilities. The introduction of disruptive technologies can lead to the need to reconsider established business models, creating uncertainty and resistance to change. Digitization raises data security and privacy issues, requiring robust solutions to protect sensitive information. Automation and artificial intelligence raise questions about the impact on labor dynamics and social equity.

This book aims to explore these opportunities and challenges in detail. Through a combination of in-depth analysis, case studies, and expert perspectives, we will examine how innovative technologies are redefining key sectors of the economy, such as health, energy, transportation, education, and much more. We will also explore the social, ethical, and legal implications of emerging technologies in order to provide a holistic perspective on technological innovation.

In conclusion, the book offers valuable guidance for anyone interested in understanding technological innovation and digitization, not just as isolated phenomena, but as forces shaping our world. Whether you are an entrepreneur, an industry professional, a student, or simply curious, you will find in these pages a wealth of information and reflections to successfully navigate the ever-changing world of innovative technologies.

Reading this book offers the opportunity to broaden one's knowledge of innovative technologies, understand their implications, and explore the opportunities these transformations offer.

We hope that this book will provide useful resources, ideas, techniques, and methods for further research on these issues.

**Fabio De Felice**
**Antonella Petrillo**
*Department of Engineering*
*University of Naples "Parthenope"*
*Naples, Italy*

# Authors

**Fabio De Felice** is a Professor in the Department of Engineering at the University of Naples "Parthenope", where he teaches industrial plants. He is a careful observer of the dynamics of entrepreneurial growth in the digital economy. Dr. De Felice is the Founder and President of Protom SpA and various companies that operate with constant attention to technological innovation. Since 2016, he has been a member of the Italian delegation to the B20 Taskforces/Cross-thematic Digitalization and SMEs. Dr. De Felice is the author of more than 150 publications in international journals and various texts on the optimization and innovation of industrial processes.

**Antonella Petrillo** is a Professor in the Department of Engineering at the University of Naples "Parthenope". She is a member of the UNI/CT 519 Commission Enabling Technologies for Industry 4.0 of UNINFO. Dr. Petrillo is the author of more than 100 international publications on issues related to the optimization and innovation of production systems.

# 1 Industry 4.0 Today, Industry 5.0 Tomorrow

## 1.1 HISTORY OF PRODUCTION SYSTEMS

The evolution of production systems is an ongoing process that has undergone **significant transformations** throughout history. These changes have been influenced by various factors, including technological innovation, economic, social, and environmental changes, as well as market dynamics. In this brief discussion, we will explore the key stages of the evolution of production systems and the driving factors behind these changes.

One of the earliest known production systems is **artisanal production**, which characterized early forms of human production. In this system, the artisan performed all stages of the production process, from design to the creation of the final product. The artisan possessed specialized knowledge and specific skills that allowed for the creation of high-quality products, but production was limited in scale and individual capacity.

With the advent of the *Industrial Revolution* in the 18th century, there was a radical transformation in production systems. The introduction of machinery and mechanical technologies made large-scale production possible and gradually replaced manual labor with machine work. This led to the era of factories and assembly lines, where workers performed specific and repetitive tasks along the production line.

The 20th century witnessed further developments in the evolution of production systems, with the introduction of concepts such as **Fordism** and **Toyotaism**. Fordism, associated with Henry Ford, relied on mass production and the standardization of products. This helped reduce production costs and made consumer goods more accessible to the masses.

On the other hand, Toyotaism, developed by the Japanese automaker Toyota, introduced the idea of *just-in-time* production and placed a central focus on waste elimination and process optimization. Toyotaism also promoted worker involvement in seeking continuous improvements and identifying solutions to production problems.

In recent decades, the evolution of production systems has been strongly influenced by the emergence of information and communication technologies. The integration of computers, robotics, and automation has ushered in a new era of digital production. Production processes have become increasingly automated, with intelligent machines and systems carrying out complex tasks quickly and efficiently.

Another important aspect of the evolution of production systems is the growing attention to **environmental sustainability**. Modern production systems seek to minimize environmental impact through the adoption of sustainable practices and

DOI: 10.1201/b22968-1

the use of clean technologies. Circular production systems are being developed, where materials are reused and recycled, reducing reliance on natural resources and limiting waste.

Furthermore, the **advancement of digitization** and the emergence of artificial intelligence technologies are opening new perspectives in the evolution of production systems. Artificial intelligence can enhance the efficiency and optimization of production processes, allowing for greater customization of products and services.

In conclusion, the evolution of production systems has been characterized by significant transformations throughout history. From craftsmanship to the era of factories, from mass production to Toyotaism and digital production, production systems have adapted to technological, economic, and social changes. The current landscape is characterized by digitization, automation, and environmental sustainability, while artificial intelligence continues to offer new opportunities and challenges. The evolution of production systems remains a dynamic and ever-changing process, driven by innovation and the need to adapt to the changing needs of society.

## 1.2　PRODUCTION SYSTEMS AND ECONOMIC SYSTEMS

An **economic system** is made up of all the **actors** who carry out activities aimed at procuring the necessary means to satisfy their needs. Many "systems" have undergone and are undergoing profound changes in recent decades due to the advent of automation and digitization, or rather their widespread diffusion in the industrial sector and beyond. We are witnessing what is called the **Fourth Industrial Revolution** or **Industry 4.0** or even **Smart Manufacturing**. However, we are already starting to talk about **Industry 5.0**, albeit with different meanings. From what we can imagine, the Fifth Industrial Revolution will be characterized by the presence of new technologies that combine the *physical, digital,* and *biological spheres.* It will generate opportunities and challenges in all disciplines and in all economic and productive sectors. We will witness the integration of technologies already used in the past (big data, cloud, robots, 3D printing, simulation, etc.), further "enhanced" because they are connected to an intelligent network capable of transmitting digital data at high speed. In other words, a *new industrial paradigm* is being born. It will generate significant changes in the way of conceiving work and man. In this new scenario, empowering **people**, or rather the importance of man in automated processes, represents a strategic factor both for the quality of goods and services produced and for the efficiency of production systems. The implementation of a new production system will evidently represent a huge change for companies, which already have to face large investments today. To remain competitive, one cannot remain indifferent to this epochal change. It is not possible to miss this opportunity. However, past and recent history teaches us that there are unforeseen rupture events that, in retrospect, are improperly rationalized and judged predictable. We refer to those unexpected events of great importance and with great consequences. **Would you have foreseen the diffusion of the internet? Or economic crises or pandemics? We often hear the "black swan"** metaphor used in these cases. But this is not the meaning thought by Nassim Nicholas Taleb[1] in his essay entitled *The Black Swan*. A "black swan" does not necessarily have a positive or negative connotation, rather it represents only an unexpected event with a strong impact on the story. Such events, considered highly divergent from the norm, collectively play a far more important role than the mass of ordinary events. It happened too with **the Coronavirus (COVID-19) pandemic.**

COVID-19 was the first major global challenge of 2020. **We are in the presence of a black swan.** Leaving China in the first days of 2020, it immediately demonstrated a high level of virality, spreading suddenly in many other countries of the world but primarily in Italy. The most important aspect to underline is that the health crisis was also characterized by a non-negligible **economic crisis at a local and global level.** It was estimated to impact at least 5 million businesses worldwide. Starting from the world of fashion to food, air transport, technology, and energy. There were many companies at risk of closure or subject to downsizing. **Ralph Lauren** closed half of its 110 stores in China, **Burberry** and **Under Armor** estimated a loss in sales of 50 million dollars. **Brooks Brothers**, the oldest chain of retail stores in America and a symbol of men's fashion, despite its more than 200-year history, could not resist the changes in the market: the brand's economic situation worsened with the consumption freeze linked to the spread of the pandemic and led the company to file for Chapter 11 bankruptcy, similar to the Italian extraordinary administration. With the news of the closure of 1,200 stores in Europe and Asia, the Spanish clothing chain **Zara** also showed that it was strongly affected by the pandemic. The company made it known that it was focusing its new growth strategy on online sales. The collapse in sales caused by the spread of the Coronavirus has also affected the well-known jeans brand **Levi's,** which laid off 700 employees, equal to 15% of the company's total workforce. The temporary store closures resulted in a 62 percent reduction in net revenue and a net loss of $364 million.

The feeling that one experiences in situations of this type, to put it in the words of psychologists, faithfully follows the **5 phases of grief** (denial, anger, negotiation, depression, and acceptance) developed by **Elisabeth Kübler-Ross**[2] in 1970. It should be noted that since this is a model in phases and not in stages, the phases can alternate and recur several times, with varying intensity and without a precise order: emotions do not follow rules, but, as they manifest themselves, they vanish, perhaps mixed and superimposed. Generally speaking, the **first phase** is that of "**denial**", in which one does not want to accept reality and tends to minimize the phenomenon or the event itself. *"It is not possible!"*, *"I can't believe it"*, and *"it can't be that this could happen to us"* are the most frequent phrases recorded in this phase. The second stage is that of "**anger**". It is here that strong emotions such as fear and despair begin to manifest. Anger can explode in all directions: it can extend to friends, family, institutions, etc. The path that first saw us condemn the Chinese, guilty of having spread the pandemic. At the end of this phase, there is the "**negotiation or bargaining**". In our opinion, this is the most important and most delicate phase since it is precisely in this phase that one wonders what can be done and which projects can be invested in. This phase corresponds to the moment of greatest **creativity**. A different world takes shape, different life models, and different businesses, and it is precisely in this phase that we have the opportunity to express the maximum of our "creative potential". If we look at examples of what has happened to the market during the pandemic period, we can remember the example of **Cosmo Lady Holdings Company Limited**, the largest underwear and lingerie company in China since 2019, based in Dongguan, which initiated a program to boost sales through WeChat by enlisting employees to promote themselves in their social circles. The company created a sales ranking among all employees (including the president and CEO), helping to motivate the rest of the staff to participate in the initiative. Essentially, most of the employees have become a sort of "influencer" contributing significantly to the promotion and sale of their company's products. Another example worth mentioning, in the early stages of the pandemic, is that of **Master Kong**, one of the main producers of instant noodles and drinks, which anticipated the phenomena of hoarding and stock depletion, and shifted attention from off-channel channels -line and large-scale distribution to O2O (online-to-offline), e-commerce, and smaller shops. By definition, crises have a highly dynamic trajectory, requiring constant reformulation of mindsets and plans. The early readiness gradually gives way to new discovery and meaning-making, then to crisis planning and response, to recovery strategy, to post-recovery strategy, and finally to reflection and learning. This process must be fast. We must be aware that the other two phases that characterize our journey (always preserving the similarity with the 5 phases of pain) are first **depression**, which we would never like to reach but which would manifest itself when all our good intentions, should they turn out to be unrealistic attempts, to then move on to the **acceptance** phase. In the digital age, technologies have not limited themselves to modifying everyday life but have significantly marked and changed our habits and our world. Without a shadow of a doubt, we can say that the pandemic has swept away many of our certainties in one fell swoop, leaving us with big questions in a very fluid magma that has not yet solidified. It is a state that start-ups know well. Visual navigation with difficulty in reaching a safe port, and forced to adapt to constantly changing scenarios. This is the world of start-ups, the one now known to all as a volatile, uncertain, complex, and ambiguous environment, the **VUCA** – Volatility, Uncertainty, Complexity, Ambiguity (Figure 1.1). In fact, it is a question of understanding how **the way of doing business must change** and what resources are needed to **effectively implement new management models.**

All organizations should become aware of the importance of knowing how to respond quickly to changes, which are generally always fast and unpredictable. Promoting a **corporate culture** that seeks **dynamic solutions and collaborations between self-organizing inter-functional teams** could help companies respond quickly to change, ensuring the flexibility that is lacking in traditional processes. No organization is immune. We must live in the awareness that the periods of economic and social crisis are not comparable to a sprint, to a hundred-meter race. Rather, they are more like an **ultramarathon**. If anything, it will be more like an **ultramarathon**. This term identifies foot races that have a distance to cover greater than 42 kilometers. Those who know this discipline are well aware of the pitfalls that lurk along the way. If we were to ask one of the athletes involved in this discipline what three suggestions could be to pass on to those who wish to venture into this path, we could summarize them as follows:

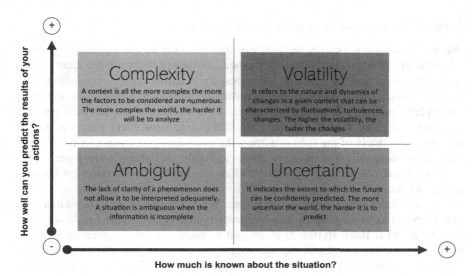

**FIGURE 1.1** VUCA (Volatility, Uncertainty, Complexity, Ambiguity).

1. **Manage resources well,** or not use metaphors. Measure out both financial and material efforts well. The management of resources in the company represents a key variable for the survival of the company in times when the road is "long" and it is not known what will happen behind "the curve". We must have enough "energy" to face the whole journey.
2. **Set clear and short-term goals.** Ultramarathon runners do not think about the final finish line as they continue their run, but about the lamppost at 700 meters or about a house visible a kilometer away. Wanting to translate the metaphor into the world of companies, they must equip themselves with short-term objectives, which can be achieved in a short period of time, so that the achievement of the same provides additional energy for the continuation of the path.
3. **Distribute "positivity pills" along the route.** During the race, the ultramarathon runners bring friends and family with them, who they "distribute" along the way to give themselves a positive emotional charge, an encouragement that facilitates the continuation of the efforts. And this is as true for individuals as it is for companies. Optimism is learned and can be learned. Precisely for this reason it is necessary to surround yourself with people who truly believe in us and our projects without hesitation and reservations. It should be remembered, as neuroscientists remind us, that we are, in some way, the sum of the 5 people we frequent most (unfortunately Facebook is also included in the calculation of the 5 "people"), and therefore we must carefully select our "travel companions", especially in the awareness that the journey will not be short.

**If this is the scenario, then what should we expect?**

We are witnessing the use of technologies that, until before the lockdown, we were told would have been impossible to implement, and which today are indispensable for our work and for this contingency. It is necessary to encourage this development and promote policies that can support innovation. Although today it is a clearly unavoidable push, this kind of political-economic orientation has known a growing consensus for years. With this in mind, the **B20 is organized every year**, the summit of the Confindustrie of the G20 countries, the privileged interlocutor of the governments on issues connected to entrepreneurship, trade, and global innovation.

Crisis situations such as COVID-19 offer the opportunity to **redesign** ways of living, producing, and doing business. We cannot think of realizing the new world and the new rules in the same way as before. We must learn to make them in a new, better, and more economically, environmentally, and socially sustainable way. We have to **Build it Back Better**. It is not just a claim. One of the main signs of the difficulties that our country is experiencing is, for example, the effort with which it is possible to *make a system*. We tend to operate individually, both as individuals and as companies, and also in the country's public administrations. Meanwhile, COVID-19 has made clear the need to think about solutions on a **global scale**. In moments like these, unity is truly strength, and it is no coincidence that consequent behaviors emerge in economic processes. To deal with the COVID-19 emergency and the restrictive measures imposed by the institutions, **virtuous mechanisms of collaboration** have also been activated between important groups and competing companies, so much so as to lead us to a new industrial phenomenon that we can define as "coopetition". Many competing companies have rethought their way of working through new **competitive-cooperative strategies** to jointly carry out one or more phases of production of a given good and/or service. In a context of crisis and scarcity of resources, no organization can have all the sufficient resources to achieve and sustainably maintain success. In this context, new and unprecedented partnerships have arisen, such as, for example, collaborations in the **pharmaceutical sector** to get treatments and vaccines sooner or the collaboration of big tech companies to find solutions to the pandemic, protect people, and allow a return to normalcy. **Apple** and **Google** have partnered to make it possible to use Bluetooth technology to help governments and health authorities contain infections, while fully respecting users' safety and privacy. Also, companies in the fashion sector made facilities and expertise available to produce basic necessities for emergencies. The **Group Prada** has started the production of 80,000 gowns and 110,000 masks for nurses and medical staff. The **Armani Group** has ordered the conversion of all its Italian production plants into the production of disposable lab coats for the individual protection of healthcare workers involved in dealing with the Coronavirus.

## 1.3   INDUSTRIAL SYSTEMS BETWEEN GLOBALIZATION AND CRISIS

The consideration is that **the darkest moments**, in reality, are moments from which new opportunities can also arise. We can say, without a shadow of denial, that it is the **perfect time to be imperfect**. As has been the case in American West Coast culture for decades, "failure" tends to be accepted as a sign of experimentation, and therefore is the best time to experiment. All of this provides an extremely fertile **environment for new ideas.** Much is changing, and some of these changes will persist in our lives (as happened after the attack of September 11, 2001); others will involve slight behavioral changes, but all will be projected toward a new normality that will permeate our lives.

We experience cyclical crises and changes that seem impossible and that distort our existence. In this regard, we are reminded of a statement by the American futurist **Alvin Tofler**[3] from 1964. According to Tofler, humanity in those days was *shocked* by the future. The great scholar identified the main reasons for this shock as too much information and excessive technological acceleration. According to the *Harvard Business Review,* our generation is also bewildered by **infoabulimia** (excess of information data) and by technological acceleration. Exactly what Tofler said in 1964. According to **Prof. Otto Sharmer**, professor at MIT's Sloan School of Management and co-founder of the Pre-sencing Institute, our era, at least until the Coronavirus, has been characterized by three "disconnections" that characterized it:

- the first: **disconnection with nature**. The first jobs related to primary production were completely in the open air in contact with nature. Then, in the first industrial revolution, the processes began to lead man into the factories and to move away more and more from nature up to the present day, where you live in companies or mega buildings far away from nature;
- the second: **disconnection from ourselves**. Always retracing the evolution of industrial production, we went from being the center of the world of work, everything depended on our work, to **George's factory Orwell,**[4] where to be effective and recognized by society you had to be part of a mechanism, almost anonymous and disconnected from the others;
- the third: **disconnection from others**. An emblematic example of this disconnection can be found in the multistory offices equipped with thousands of workstations, each separated from the other by screens or partitions aimed at generating a functional separation between the various operators.

We must take advantage of this moment and the technology that is offered to us, to eliminate these disconnects. We have never been so much connected to the network as disconnected from each other, as **Simon Sinek** reminds us,[5] this phenomenon excessively affects our young people (millennials and generation Z in particular).

**How to do?** We have **solutions at hand**, but often we do not have the key to understanding them. We explain this better with an example. In the 19th century, aluminum was considered the **"metal of kings"**, more precious than gold and platinum, so much so that it is said that it was chosen by Napoleon III to amaze the king of Siam (now Thailand). Thus, the emperor of France, who always used silver in the palace, decided to set the table with aluminum plates and cutlery. In fact, if you think about it, and as Charles Dickens (1857) recalls, aluminum is a white metal like silver, unalterable like gold, easy to melt like copper, hard like steel, malleable, ductile, and has the unique property of being brighter than glass. Aluminum is present in large quantities on earth, but rarely in free form, it makes up about 8.3% (by weight) of the earth's crust combined with other elements (mainly sulfur, silicon, and oxygen). At that time the method of direct reduction by electrolysis proposed in 1854 by Claire Deville was not yet known,[6] hence its preciousness. Once the way to extract aluminum was discovered, given its great availability in nature, its use grew, and today we know that we also use it in multiple ways. This example wants to offer all of us food for thought. Today we speak of scarcity of resources, for example, energy, while we know that we are on a planet surrounded by energy, solar energy, 5,000 times greater than the one we use in a year. Every 88 minutes, 16 terawatts (TW) of energy hits the Earth's surface. Therefore, as in the case of aluminum, we cannot speak of a scarcity of resources, since there are plenty of them, but of **the accessibility of these resources**. We must concentrate our energies to find solutions that make the immense resources available accessible and therefore usable, we must find our "electrolytic" process that provides us with the key to open the door to the abundance of resources present on our earth. We could do a similar example with water. We know that we live on a planet made up of 70% water (of course 97.5% is salty, 2% is frozen, and we are "satisfied" with 0.5%). Imagine if the "electrolytic" solution were found that would allow us to have access to this enormous amount of resources. At the very least we would have solved the drought problem in Africa without thinking about the many opportunities that could be generated.

With these examples, we want to show how everything COVID-19 has led us to (as well as any other critical scenario) can and could represent an opportunity to change paths and find solutions that, until recently, we thought were unfeasible.

The solutions that we have the opportunity to experiment with can drastically affect our lives as much as our companies. The global lockdown has clearly brought out a critical issue in supply chain management, and delocalization has made the fragility of our supply chains clear. However, for every ending, there is a new beginning, another opportunity, a new strength. Technology helps to develop innovative solutions that favor the recovery of economic, social, and productive systems. In this regard, the Commissioner for Innovation, Research, Culture, Education and Youth, **Mariya Gabriel**, said that a time of crisis requires resources to be channeled toward achieving rapid results, guaranteeing support funding for the completion of a connected digital global market and support for the development of creative industries and a successful European industry.

Progress continues to travel with the steps indicated by **Moore's law,** and in our opinion, the acceleration imposed by COVID-19 could override the geometric progression proposed by the co-founder of INTEL. As is known, in 1965 **Gordon Earle Moore**[7] hypothesized that the number of transistors in microprocessors would double approximately every 12 months. Indeed, in 1958, the first integrated circuit provided for the presence of only two transistors. In 1971 the number of transistors increased to 2,300, and in 2016 this number reached 14.4 billion in the Intel i7 processor.

A few years later, **Ray Kurzweil**, an American inventor and computer scientist, demonstrated that this law applies to any tool that becomes "computer technology".

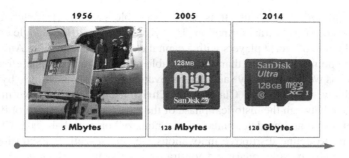

**FIGURE 1.2** Evolution of the hard drive.

Therefore, Moore's law has become the yardstick and goal of all companies operating in the hi-tech and new economies such as Intel or AMD, Google and Amazon, and Microsoft or IBM. We recall that the first IBM hard drive, **IBM 305 RAMAC** (Random Access Method of Accounting and Control) from 1956, not very easy to move from one office to another (see Figure 1.2) with a power of 5 Mbytes, has become the 2005 in a **128 Mbytes micro SD** at a cost of 99 dollars (against 120,000 for IBM), then arriving in 2014 at a **128 Gbytes micro SD** at the same cost.

Technology runs fast, faster than our imagination sometimes does. You probably remember the smart lenses worn by Tom Cruise co-star Jeremy Renner in the Ghost Protocol episode of **Mission Impossible**. Well, Google, in 2014, started the **GOogle Contact Lens project** a new technology seen in the new chapter of **Mission Impossible**. The feeling is that we can continue to imagine that our future follows Moore's law. All this leads us to affirm, in agreement with Peter Diamandis and his most famous partner, **Elon Musk,** CEO of the futuristic Tesla as well as founder of **Space Exploration Technologies Corporation** (SpaceX), a US aerospace company established in 2002 with the aim of reducing the costs of access to space and allowing the **colonization of Mars,** that it will not be technology or money that will solve humanity's great challenges, but a determined and motivated human mind with a great passion and a clear goal.

## 1.4 DIGITAL REVOLUTION: AN EPOCHAL CHANGE

Historian **Paul Kennedy** in his *Rise and Decline of the Great Powers* shows in detail the dynamics that lead nations to increase their power and then to lose it. In these scenarios, **industrial revolutions** often play the role of engine of change: **those who excel in technology also excel on the economic and geopolitical front**. Those who have managed to excel in the development of the latest technologies available have in the past obtained an advantage over other players. When, in 2016, **Google** reached 395.2 billion in capitalization, a figure equal to 3.1 times the turnover accrued during that year, it was therefore immediately clear that the world was witnessing a key moment in the advance of the **Digital Revolution**, the phenomenon that in the last 20 years has redesigned the world scenarios not only at an economic level but also at a social level, affecting the way we live, work, and communicate for each of us. It is an unstoppable process characterized by an unprecedented speed of transformation. Two years later, in 2018, Google entered the very small circle of world players with a capital exceeding one trillion. And the news was, at that point, little more than an inevitable confirmation. It is therefore evident once again, as we have already said, that innovation manifests itself little by little and then, suddenly, very quickly. **Clayton M. Christensen**'s[8] studies have marked the clear passage between the historical phase of the experimental research of innovation and that of a true and modeled innovation "industry" on a global scale. Christensen was the first to define disruptive innovations, i.e. the **so-called disruption**, distinguishing them from evolutionary and non-transformative innovations, a concept that is essential today in any industry and sector. What is a disruptive innovation? To get an idea, just look at the last 20 years. Much has changed in the existence of human beings and businesses since, 25 years ago, the scholar defined the foundations of innovation in the economy and its adoption in society.

*What has happened in the last twenty years? How has the world around us changed? How have we changed, our habits, our way of interfacing with it?*

Twenty years ago the economic markets were dominated by big brands such as **Coca-Cola** and **McDonald's**, producers of consumer goods and brands with global recognition, supported by players such as **IBM, Cisco,** and **HP**, companies with a great innovative drive but focused on the production of hardware.

The first two decades of the 21st century were marked by the rise of the big names in the digital world, service providers who have ousted the producers of goods from the rankings of the biggest global brands.

It is the affirmation of the supremacy of the *intangibles*, the maximum expression of an increasingly disintermediated economy, whose protagonists are service providers in most cases totally devoid of material infrastructure: from **Airbnb** to **Uber,** from **LinkedIn** to **Facebook**. Value is dematerialized. Uber does not need cars to be the largest taxi company in the world, or Airbnb does not need physical accommodation facilities to position yourself at the top of the rankings of the largest hotel chains. These are players who, due to their size and level of innovation, have had, in just a few years, the strength to sweep away the previous economic scenario, redesigning a new one, which, through the implementation of new digital technologies, is capable of building revolutionary business models that transform the management of products, services, and information.

It is the era of **digital plutocracy**, a new socio-economic dimension in which, according to a Pareto proportion, a few giants share large slices of the market – **Amazon, Google, Microsoft, Facebook,** and **Alibaba** – are the new economic oligarchy, innovation outpost, and funnel within which small and medium-sized competitors are sucked in, with a force that appears unstoppable.

The power of these giants does not lie in their ability to produce more, but in their ability to offer consumers exactly what they need.

The revolution in production relations is one of the most disruptive aspects of digital transformation: in fact, there is a revolution of production relations: from the unidirectionality of mass production, we move on to a new mode of interaction between producer and consumer, characterized by a constant circular process that sees the response to the need to be immediately followed by feedback from the customer, which becomes the starting point for a refocusing of the output, according to the indications received. It is in this continuous succession of feedback and *re-targeting of the need that the so-called mass customization* takes place, the production tailored to the customer's requests.

In this sense, Alvin Toeffler coined the term prosumer in 1980, suggesting a possible fusion of the roles of producer and consumer. The "Prosumer" is an active protagonist of the process that involves the creation, production, distribution and consumption phases. Today this role is emphasized by the explosion of Social Media and Social Networks, real catalysts of change with regards to changes in purchasing experiences.[9] Today it is understood that being able to follow the product/service in the phase of its life following the sale allows us to retrieve valuable information **from and about the customer.**

*Digital technology allows for the construction of a new relationship with the customer. Everything becomes interactive. The needs of consumers are better understood, and their retention can be improved.* It was essential to introduce *after-sales* as a marketing strategy in his companies in order to offer a customer service that does not end with the sale and allows, for example, active maintenance with a remote contact, without the direct intervention of the operator, but with a sending of instructions sufficient to repair the machinery, without considering the enormous amount of information that this process allows to return on the consumer and on the use of the acquired good. Before the digital age, after-sales was a relational service functional to maintaining the relationship with the customer. Today it is an expression of *added value*. However, when a "phenomenon" such as the COVID-19 emergency occurs, which interrupts the transmission of information, it is not clear how the consumer will behave, and if there is a structure such as the Italian one, oriented toward B2B, everything becomes much more complicated because there is a very long and complex supply chain of suppliers and sub-suppliers. However, it is good not to draw conclusions easily because the complexity is enormous and it is still difficult to understand the real change in consumer habits. In any case, we have one certainty. The daily overdose of information has profoundly changed the user's behaviors, needs, and relationship with brands and, ultimately, the entire *customer journey*. It is a new type of relationship based on *ask and advocate* (research and word of mouth); the online dimension, in fact, allows the customer to search directly on the net for information about a brand from those who knew and bought it before him.

It is therefore possible to state that, if in the pre-connectivity era, the individual customer formed his own independent opinion on a brand, today the same is strongly influenced by the community surrounding the customer. The weight of the community is also the reason why it is no longer possible to identify a customer's loyalty to a brand through the repeated purchase of the same; what matters is the customer's willingness to recommend it to other people.

Along the lines of the customer journey proposed by **Philip Kotler**, therefore, we can say that today the consumer in his purchases follows the 5-A model: Aware, Appeal, Ask, Act, and Advocate.

The first phase is the one in which the consumer is passively exposed to a wide range of information, becoming aware of the existence of a large number of brands. It is a discovery phase in which word of mouth and brand marketing allow the (potential) customer to get an idea of the values and elements of differentiation of each brand. It will be this last aspect that will play a key role in attracting (appealing) the attention of the customer, who is constantly looking for the "**wow factor**". Amazement also induces the customer to enter the pro-active phase of the customer journey, prompting him to seek information about the brand (ask). The digital age means that, in this phase, unlike in the past, a fundamental role is played by the initiation of a social path, research whose most significant results are not those deriving directly from the brand but those from conversations with other people who have already had experiences with the "brand". Only at this point does the customer actually take action (act) not only through the purchase of the product, but above all through its use and interaction with the after-sales services. Therefore, the better this experience will be, the higher the opportunity for retention, i.e. for repeated pur-chases, which derives from the establishment of a strong sense of belonging to the brand and translates into the recommendation of the product to others (advocacy).

Ultimately, the association between purchase and experience is strengthened: *choosing a brand means adhering to the values it transmits to the community that supports it.*

Object of transformation in the digital society, therefore, is also the **perception of value**, which is transferred from the good to the benefit. The spread of **platforms** should be interpreted from this point of view as an innovative means of creating, distributing, and providing these benefits, the embodiment of an unprecedented paradox: unlike goods, subject to wear and tear such that their value decreases with each use, the platforms stand out for reversing this proportion, seeing their value increase the more their use increases.

And it is, however, again by virtue of the unprecedented experiential nature of the purchase that explains the phenomenon that is only apparently antithetical to the affirmation of the primacy of the *"intangibles"*. In order to remain lasting and solid, the link between the brand and customer needs to get out of the digital dimension, move into the real world, interpreting it no longer as an access point for purchases, but as a necessary integration of a 360-degree experience. The leasing operation of **Netflix**, the giant in the production and distribution of entertainment content in streaming, of the Paris Theater in New York should be read from this point of view. In November 2019 this historic cinema, the last single-screen New York movie house, was on the verge of closing its doors, overwhelmed by the colossal multiplexes and the unstoppable pressure of the big streaming players who in just a few years have transfigured the entire industry cinematic. It was precisely the intervention of Netflix that saved it, which managed to lease it, making it the location for the premiere of its production, which later won multiple Oscars, *Marriage Story*. The Paris Theater therefore becomes, for Netflix, the exclusive venue for premieres and special events, which extend the life of the platform beyond the screens of smart TVs and PCs.

Even more poignant is the case of Amazon, the undisputed hegemon of *digital retail* and the largest Internet company in the world which, in 2016, created **Amazon Go**, the first "**Just Walk Out Shop**" in Seattle. By bringing the distinctive technologies of its platform into a physical space and integrating them with *features* developed for self-driving cars, such as computer vision, sensory fusion, and deep learning, Amazon Go creates an innovative shopping experience in which, simply, you enter the store, choose the desired products, and exit, without queuing and without paying at the cash desk, but by direct debit to your Amazon account. The success of the experiment carried out inside a relatively small store (167 m$^2$), prompted the US giant to broaden its horizons: in 2017, with a 13.7 billion dollar operation, acquired **Whole Foods**, the most popular organic food chain in the United States, with 460 stores and a leadership position in the fresh food market. 2020 is the year of the qualitative leap: **Bezos** launches, again in Seattle, **Amazon Go Grocery**, a supermarket that capitalizes on the technological experience of the "younger brother" Amazon Go and carries it on a large surface area (about 1,000 m$^2$), also opening up to fresh goods, thanks to the leadership gained in the sector thanks to the acquisition of Whole Foods (we look forward to hearing the results of the recent experimentation of Amazon One, the new frontier of purchasing with biometric data, by scanning the palm of the customer connected directly to a credit card selected by the customer).

The cases of Netflix and Amazon make it clear that, contrary to common thought, the role of technology in the era of the Digital Revolution is not to create valuable **experiences but to make them possible** through the identification, implementation, and management of unprecedented channels and forms of contact between producers and consumers. The effectiveness of these activities will be directly proportional to the ability of the subjects in the field to dominate and anticipate change, which constitutes the fundamental pillar of success in the age of digital transformation.

As **John Steinbeck wrote,**[10] *Once the miracle of creation has started, then the group can intervene and enhance it.* A statement that seems to adapt perfectly to the scenarios of this new era, which is not characterized by great inventions but by the ability to accelerate and strengthen that arises from the integration of technologies, know-how, knowledge, information, and which is all the more effective the more it is shared. Ultimately, it is a question of learning to read reality in an unprecedented, lateral way, observing the weak signals and starting from them to build new ways of interacting, creating, producing, and living.

## 1.5 SMART FACTORY AND INDUSTRY 4.0

**Alvin Toeffler** stated: *The illiterates of the 21st century are not those who cannot read or write, but those who are unable to learn, unlearn and relearn.* The characteristic of the new paradigm in which we are immersed is its circularity, the perpetual departure from what it was in order to then return to reinterpret it: the dematerialization that becomes experience, the primacy of the machine that exists only if at the service of a human purpose, the production autonomous and intelligent that allows to optimize the integration of the human component in the production system. It is in this scenario that the key role of data, of their collection, of their management, but above all of their interpretation is played out. Defined as the **new oil**, they require new "refineries": interpretative methodologies that allow them to be translated into information useful for consolidating the ability to anticipate change. This is a crucial issue, which has implications for all areas of human existence, but which has played a particularly important role in the advent of the so-called **Industry 4.0**. Industry 4.0, far from being identified with a new technology or a set of new technologies, manifests itself as the possibility of combining and integrating existing technologies, whose cost has decreased ever more significantly, increasing their ease of use. It is in this sense that it can be said that Industry 4.0 is a **consequence** of the very spirit of the digital world. The digital world is in fact dominated by interoperability: a product created for one platform can be adapted to another, become another service, and start the carousel all over again. Computers, for example, what are they for? To nothing precise, they can do anything by representing the world in the form of bits. The same goes for the internet, where almost anything can become a multipurpose engine: a tool that allows the construction of other tools. Think of **GitHub**, the world's largest community of developers, a sort of virtual warehouse where anyone who develops codes can take, leave, and share them. GitHub currently hosts 28 million developers and nearly 100 million projects. What is GitHub if not an immense network factory built on the denial of Fordist linearity? Just as the 20th-century factory drew its strength from its ability to anticipate, the factories of the new world feed on fruitful unpredictability. Today the future is non-linear, and the so-called Industry 4.0 is basically a consequence of it.

A direct emanation of Industry 4.0 is the **Smart Factory**, which is defined as such not by virtue of a mere strengthening of robotization and automation practices, but by its ability to communicate and interact. A reorganization that makes it possible to respond to the constant evolution of demand. It exploded in **2011** in Germany and from there extended to the United States of America and Europe. The Fourth Industrial Revolution outlines a new production scenario, made up of smart factories *in which all systems are able to communicate and interact in real time, thanks to a network called the Internet of Things able to simplify and rationalize work.* On the one hand, enabling technologies, from Advanced and Additive Manufacturing solutions to Virtual and Augmented Reality, up to Cloud, Cybersecurity, and Big Data technologies, passing through vertical and horizontal integration systems along the entire value chain. On the other hand, a production reality forced by the times to seize the opportunity to produce in a new, more effective, and efficient way, customized to the needs of the individual and capable of improving the integration of the human component into the production system. The result is a system that as a whole is faster, and, more flexible, and at the same time capable of ensuring quality, making it possible to increase productivity and, consequently, competitiveness.

## NOTES

1 Nassim Nicholas Taleb (Amioun, 1 January 1960) is a Lebanese philosopher, essayist, mathematician, and academic naturalized American, an expert in financial mathematics and probability theory.
2 Elisabeth Kübler-Ross (8 July 1926, Zurich–24 August 2004, Scottsdale) was a Swiss psychiatrist. She is considered the founder of psychothanology and one of the best-known exponents of death studies.
3 Alvin Toffler (New York, October 3, 1928–Los Angeles, June 27, 2016 [1]) was an American essayist, who defined himself as a futurist. For many years he has studied the mass media and their impact on the social structure and on the world of culture.

4  George Orwell (25 June 1903, Motihari–21 January 1950, London) was a British writer, journalist, essayist, activist, and literary critic.

5  Simon Oliver Sinek (London, October 9, 1973) is an English naturalized American writer and essayist. He is the author of several books on communication and leadership topics, including the best-sellers Start With Why and Leaders Eat Last.

6  Henri Étienne Sainte-Claire Deville (Saint Thomas, March 11, 1818–Boulogne-Billancourt, July 1, 1881) was a French chemist, known for his research on aluminum.

7  Gordon Earle Moore (born January 3, 1929) is an American entrepreneur and computer scientist, co-founder of Fairchild Semiconductor in 1957 and Intel in 1968.

8  Clayton Christensen, a Harvard professor, was the scholar who best defined the foundations of innovation in the economy and its adoption in society and theorist of "Disruptive Innovation", or theory of "disruptive innovation", defined as the entrepreneurial idea most influential of the early 21st century.

9  Alberto Baban, Alberto Mattiello and Armando Cirrincione. Mind the change. Understanding change to design the business of the future, Guerini Next, 2017.

10 John Ernst Steinbeck, Jr. (Salinas, February 27, 1902–New York, December 20, 1968) was an American writer among the best known of the twentieth century, the author of numerous novels, short stories, and short stories. He was briefly a journalist and war reporter in World War II. In 1962 he was awarded the Nobel Prize in literature with the following motivation: "For his realistic and imaginative writings, combining sensitive humor and acute social perception".

# 2 Enabling Technologies

## 2.1 ENABLING TECHNOLOGIES

In recent years, in most sectors, business models are changing radically, and the elements underlying their success are changing. This change is generated by the pervasiveness in the economic and social systems of new distinct phenomena, among which the phenomenon of the digital economy is becoming increasingly important. **Digital technologies** and the **logic of sharing** are favoring the birth of business models and consequently, completely new companies, tying existing ones to a radical change and/or update in order not to find themselves in a disadvantageous position compared to their competitors. Companies find themselves facing a new era that is taking them step by step to a real digital transformation. In this context, **enabling technologies** or KETs (from the English *Key Enabling Technologies*) play a fundamental role. KETs represent an opportunity for economic growth as they "revitalize" the production system. According to the definition given by the **European Commission,** enabling technologies *are knowledge-intensive technologies associated with high R&D intensity, rapid innovation cycles, substantial investment expenditure, and highly skilled jobs.*

The enabling technologies are diverse and of different types, ranging from robotics to cybersecurity, the cloud, and software. Their role is decisive in the processes related to digital transformation, as they are considered **drivers of innovation**. Generally, **nine enabling technologies** are classified. However, it is worth emphasizing that these are constantly evolving systems and that they often interact with each other or integrate with each other. Therefore, our goal is to analyze only the most salient aspects of each technology.

## 2.2 AUGMENTED REALITY, VIRTUAL REALITY, AND MIXED REALITY

In introducing the first technology, we are reminded of **Albert Einstein's famous phrase:** *Reality is a mere illusion, albeit a very persistent one.* Well, the first step to take in approaching these technologies is to distinguish between virtual reality (VR) and augmented reality (AR). When we talk about VR we are referring to a three-dimensional surrounding environment, not real but simulated, in which the user is able to interact thanks to the combination of hardware and software devices that offer a totally immersive experience. Immersiveness is among the most substantial differences with AR, but it is also what makes VR. It is possible to view VR through special headsets, such as the Oculus Rift. Other VR headsets use your phone and VR apps, such as Google Cardboard or Daydream View. While, it is called AR a technology that exploits the displays of mobile devices, wearable

DOI: 10.1201/b22968-2

devices, vehicle windows, and interactive windows to add information to what we see. Ultimately, when we talk about augmented reality, we are referring to our natural field of vision which can be enriched with new information thanks to a tablet, a smartphone, or special glasses. If we refer instead to VR, we need to imagine a completely reconstructed world in which we can immerse ourselves completely by wearing completely closed visors. **AR and VR are inverse reflections of each other:** VR offers a digital recreation of a real-life environment, while AR provides virtual elements overlaid with the real world.

AR technology is not young, so much so that the first devices were developed as early as 1968 by **Ivan Sutherland,**[1] and isolated applications of AR have been recorded since the early 90s, thanks to the diffusion of portable devices, Internet, and GPS technology. But it is with I4.0 that AR technology makes its full potential available. In fact, **AR is a form of visual content management 2.0**. There are several uses for both AR and VR. Many companies have started using AR to promote their products or services. Among the retail brands that have been pioneers in using AR are **Lego, Tesco,** and **Ikea**. AR plays an increasingly important role in e-Commerce. At the end of the 90s, both the processors capable of processing the images and the devices available were at their first experiments. Thus, all the experiments carried out were of mere little use, except in terms of advertising or scenographic for the mass market and for companies. In the first decade of 2000, **Apple** introduced the world's largest platform for AR on *mobile devices*. At that point, as often happens, the experiments begin to take on some form of utility. There are several applications that are born to experiment in the world of AR, more and more preformed, more and more uninteresting, and more and more "useful" in the different worlds. An example that straddles medicine and teaching can be found in an application that scientifically investigates the behavior and composition of the human body and its organs. **Complete Anatomy** is a very interesting iOS app that allows the "scientific" study of the human body. Thanks to the Motion Capture technology and the LiDAR scanner of iPad Pro, physiotherapists and patients will soon be able to precisely quantify the progress made in the course of motor rehabilitation as well as disseminate innovative tools in the world of DSA disorders.

An experiment born from the collaboration with our groups of programmers/ researchers and various rehabilitation centers is **RiLab "Laboratory for Rehabilitation",** an open and extremely flexible framework. RiLab is an immersive reality solution for the rehabilitation of patients suffering from neuro-psychomotor pathologies that uses virtual stimulation to induce physical recovery through complete visual and auditory immersion without the aid of any wearable device that limits or alters the freedom of interaction. The goal is to create virtual, immersive interactive 3D environments that can be projected onto any type of horizontal and vertical surface, and with which the patient interacts through simple movements, thanks to a system based on some basic components (standard PC + projector/video wall + comprehensive license e.g. basic + additional exercises) completed by the use of a large, articulated and modular set of sensors and actuators:

- With **biometric sensors,** it is possible to track movements of the whole body without the patient wearing or holding anything.

- With **laser-IR sensors,** it is possible to manage tactile interaction with any surface of any size and orientation with high precision.
- With **pressure sensors** (load cells), the user can interact with VR simply by standing or even sitting.

Solutions such as RiLab are part of "hybrid" systems called **mixed reality** (MR) or when we witness the "mixing" of digital and virtual worlds to produce new environments and visualizations in which virtual and physical objects coexist and interact with each other in real time. In other words, hybrid reality is a real mix of physical and VR. Well, through a system of sensors and actuators that can be integrated, RiLab stimulates operations of matching **objects-colors** (with or without dragging), matching **objects-dimensions** (with or without dragging), **memory musical notes**, **place-object memory** (with or without dragging), and others (Figure 2.1).

VR can be used to intervene in impaired psychic functions or on the execution of motor activities. It can also be a valuable tool for intervening on more global aspects related to the well-being, degree of participation, and autonomy of the person with a disability. Never before has the coexistence of innovations in technologies, software, and models of their use changed our lives even more radically, introducing and involving other senses and permeating the objects and habits of our daily lives. But if we look at the use of these technologies in all companies, one of the main objectives is certainly training, where these tools are used, for example, as remote support, but also and above all to carry out simulation activities. **Walmart**, the largest chain of large-scale retail trade founded by **Sam Walton** in 1962, was among the first giants of the old economy to introduce VR to train and increase the skills of its collaborators. In fact, Walmart, operating in a particularly difficult sector, has focused heavily on the reconversion and reskilling of its collaborators and having a widespread distribution throughout the world, has decided to use AR/VR technologies for training by investing around 4 million dollars alone in hardware to buy 17,000 oculus and test this new form on around one million people, distributed across its 4,000 stores. At the same time it generates enormous economies of scale as well as the possibility of remotizing different activities and

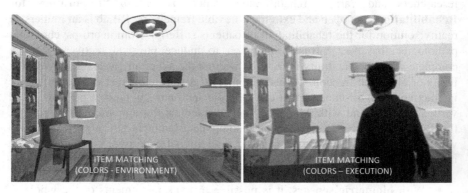

**FIGURE 2.1**   RiLab application examples.

managing them all from their headquarters. But the GDO is not the only pioneering sector in this field; the manufacturing industry has also ventured into the use of AR technologies. **Candy Hoover Group**, the historic Italian manufacturer of household appliances founded in 1946 in Monza by Eden Fumagalli and only recently passed into the hands of the Chinese Haier, in 2016 in the most important appliance fair in Germany, offered its customers an immersive experience that, taking advantage of the by now acquired complexities of gaming (think of the various playstations and XBOXes), led the customer on a journey into the house of the future, discovering and interacting with the different appliances. A **shortage** of these experiences to date is determined by *poor brain signaling of feedback* that does not associate reality and fantasy. To deal with this problem and improve the usefulness of these instruments, many studies are being carried out on **haptic sensors**. The haptic interface is a device that allows you to receive tactile sensations in response to a signal (feedback or feedback).

Since the last century, research has gone toward the discovery of technologies capable of creating completely convincing artificial stimuli for some senses: think of the high fidelity of audio reproduction or the high definition of television screens and computer graphics of video games more modern. Today research is also pushing the attempt to stimulate other senses. The dynamic nature of tactile perception means that when we use it to learn about the world around us, we do so by actively exploring the objects that make it up. The difference with senses such as smell, taste, or hearing – which are like windows open onto the world through which we receive stimuli in an almost totally passive way – is enormous. Even sight, which also makes use of various important active functions, such as gaze direction and focus, is less linked to the sensory-motor structures of our body and to interaction with the environment. Man tends to create a sort of image through touch. In order to achieve this goal, he therefore touches, skims, feels, follows, and presses the object of his investigation: in doing this, he mainly uses his hand. The hand, with the very acute sensitivity of tens of thousands of tactile receptors in its skin, is to be considered the true organ of touch, as far as higher cognitive functions are concerned. Haptic sensors are a fundamental ingredient of any VR system that seeks to provide a compelling sensation of immersion, of being present in environments in which we are not. Imagine the usefulness of these solutions for training a surgeon, for simulating driving a vehicle, or for entertainment and education. Moreover, current VR techniques are still mainly based on the sense of vision alone, creating images on optical displays, which today reach very high resolutions and almost photographic realism with excellent three-dimensional rendering. Only with the integration of other sensory modalities, in particular tactile, will it be possible for these systems to make a leap forward, thus allowing them to pass from the experience of being present in a virtual environment to that of coming into contact with that environment.

**So how can we create devices that allow our hands to generate what a projector, or a conventional computer screen, produces for our eyes?**

The science and technology of haptic interfaces (from the Greek "to touch") are dedicated to the study and creation of images by hand. Haptic interfaces are machines that allow you to generate tactile stimuli, touch, feel, manipulate, alter, and create virtual objects. However, despite the many technological advances in the field of haptic interfaces in recent years, the goal of developing fully functional and convincing touch displays has not yet been achieved.

The greatest diffusion of MR and AR solutions occurred with the development of applications on mobile devices. An interesting example in our opinion, is offered to us by **iScape.** This app allows you to visualize and test new ideas for outdoor spaces in your home before putting them to work, choosing from hundreds of different plants, and collaborating with friends. The same idea but functionalized and oriented toward a new shopping experience has been taken up by IKEA. **IKEA Place** is the way chosen by the furniture giant to support customers in making the most suitable choice for their home, browsing the thousands of items offered by the Swedish company or seeing the final result. Not only is it possible to "find yourself in 360°" in a fully furnished room, but with this app we try to offer the opportunity to customize it according to your preferences. In the world of edutainment, an interesting solution is proposed by **JigSpace**, an app that allows to explore the inside of objects in an interactive and three-dimensional way as well as create emotional presentations or prepare and use AR instructions to disassemble or assemble everyday objects. Today similar solutions are already included in the software of the new Apple iPhone. However, we are sure that we are only at the beginning of an epochal change that AR will bring to our lives. In fact, AR has also entered the luxury sector. **Gucci,** the well-known Italian fashion house active in the high fashion and luxury items sectors, recently launched an app that allows users to virtually try on shoes thanks to AR applications. It should be noted that AR technologies lend themselves to many sectors: from classroom education to museums, from medical simulations to gaming, from advertising to shopping. We find AR applications not only on devices such as smartphones, iPhones, tablets, and iPads, but more and more, also for industrial solutions, we can find wearable devices on the market. We have already written previously about how digital giants such as Google and Samsung have patented contact lenses with AR (the aforementioned Google Contact Lens). Microsoft, with its **Hololens,** seethrough eyewear, is looking for experiences in education, industry, and everyday life. AR not only changes the user's perception and interaction with the product, but also helps innovate design, manufacturing, marketing, sales, and service. AR for industry boasts numerous applications for manufacturing. Solutions that improve employee training, provide real-time instructions for the assembly and maintenance of a component, thus increasing the level of competence, safety, and overall efficiency. AR is, therefore, the enrichment of the perception of the surrounding context with digital data generated thanks to technologies that allow the superimposition of contents (such as texts, images, live action, or animated films) perceived as part of the real environment in which the subject finds himself.

## 2.3 ADDITIVE MANUFACTURING

Additive manufacturing (AM) technologies are defined as *those processes that aggregate materials in order to create objects starting from their three-dimensional mathematical models, usually by overlapping layers and proceeding in the opposite way to what happens in subtractive processes (or removal of shavings)*. We are referring to three-dimensional printing and the techniques and technologies related to it. Historically, the birth of AM or 3D printing began in 1986, with the publication of the patent of **Chuck Hull**,[2] who invented stereolithography. Since **1986**, 3D printing has evolved and differentiated, with the introduction of new printing techniques and countless materials (materials of plastic, metallic, and polymeric origin). Thus it is becoming an important technology capable of revolutionizing traditional production models. The new frontiers of additive manufacturing will be able to change the scenario of the manufacturing industry. Let us imagine if in the near future more and more "files" would circulate instead of goods. Imagine, the port of Rotterdam where instead of large spaces for storing containers, thousands of printers were installed capable of reproducing all kinds of materials or plant and animal molecules (therefore also fruit, vegetables, and other foods). Instead of ordering "objects", we pay for a file that contains the specifications of the requested product, but made in the place we want, when we want, and at the times we request. Of course, it seems too futuristic a scenario, but how many of us would have imagined having a computer, a camera, and much more in a "mobile", which today we call a smartphone? As shown by a research conducted by **IDTechEx**, a Boston-based information services company, the global market for 3D equipment will reach 31 billion dollars by 2029. According to the study, in particular, 75% of new commercial and military vehicles will have a 3D printed engine. 25% of surgeons will practice on anatomical models that are 3D printed. We were in 2018 with **Jay Rogers**, the CEO and co-founder of Local Motors, a new generation car company that offers a particular model for the design and construction of cars by combining additive manufacturing and open innovation. Jay, an eclectic man, with his elegant bow tie welcomes us in a very informal way by presenting the open innovation platform that allows us to aggregate a community of designers and producers located in different parts of the globe. **Rally Fighter**, is the "first born" of a distributed and widespread co-production in which there are parts made with 3D printers in Africa, others in Europe and in various countries of the world. Then, parts are assembled with the help of a team of workers of Local Motors and give life, in times impossible for a traditional automotive industry, and with a level of customization still non-existent in automotive production today. The success of this experiment carried out by Jay and his collaborators saw its culmination in the episode of the Fast and Furious saga in 2017 (the Fate of the Furious).

In our various exploratory journeys in the world of new technologies, we have come across different types and applications of printed matter with AM. Certainly one of the most challenging and futuristic applications is the one linked to the **WASP** (World's Advanced Saving Project) project. Project that was born in 2012 with the aim of building a 3D printer capable of realizing, with waste materials or materials found directly in the area, entire houses at negligible costs. Thus, in 2015 the Big Delta was born, a giant printer, 12 m high and 7 m wide, equipped with different types of extruders for fluid-dense mixes. The change of state of the deposited material takes place through the evaporation of a solvent that can be of a different nature, not just the water itself (Figure 2.2). After all, if we think about it, the clay house is not a recent discovery.

**FIGURE 2.2**  BigDelta WASP 12MT.

The **house of the future** will be printed anywhere, in an economic and eco-sustainable way and can be created according to needs, and in the forms on which architects and designers can indulge themselves. A first example of an architectural exercise with 3D printers we found, almost by chance, outside the **Emirates Tower One** (also known as Emirates Office Tower) along Sheikh Zayed Road in Dubai (Figure 2.3). However, Dubai is considered a land of strong experimentation on the theme of the house of the future. In fact, the first fully functional 3D-printed office building was printed and assembled right in Dubai in just 17 days. From a study carried out by the **Emirates Technological District** it is expected that 25% of the world's buildings will be produced in an additive way within the next seven years.

AM, however, is not limited to the building sector but is expanding in every sector: from the manufacturing industry, to the goldsmith sector, to the automotive sector. The University of Maine in the United States was awarded in 2019 the Guinness World Record for the largest 3D-printed object and also for the largest 3D-printed boat. The feat was achieved by the UMaine Advanced Structures and

**FIGURE 2.3**  Dubai examples of buildings made with 3D printer.

Composites Center division, also thanks to a series of large investments and collaboration with Oak Ridge National Laboratory and several other federal agencies. The 3D-printed vessel is 7.62 m long and weighs 2,268 kg. With traditional techniques, the construction would have taken weeks or even months, but thanks to the largest 3D printer in the world, it was completed in just 3 days (https://youtu.be/34F71XqvOjg). Obviously, the interest in the results achieved is enormous. 3D printing is seen as a technology capable of supporting the local economy of one of the most heavily forested and vegetated states in the United States. This is also thanks to the use of a 3D printing and construction material obtained from cellulose, a wood derivative, for projects and applications in the civil, infrastructure, and defense fields. There are in our opinion some sectors that are benefiting and will benefit a lot from this technology. We refer essentially to the aerospace and defense sector and the medical sector. As for the aerospace and defense sector, the first 3D printer launched into orbit on September 21, 2014, was called Zero-G, and was built by NASA, in collaboration with Made in Space. Two years later, a second Made in Space 3D printer, called the Additive Manufacturing Facility (AMF) was sent as a permanent facility on the International Space Station (ISS), to provide hardware manufacturing services to both NASA and the US National Laboratory on board. But the notoriety of these solutions to the general public took place in 2014, when the Italian astronaut **Samantha Cristoforetti** created small spare parts, renouncing external procurement, through the POP3D, a portable 3D printer branded Altran and Thales Alenia Space. Of particular interest is the bilateral agreement stipulated between the Italian Space Agency (ASI) and the Russian Space Agency Roscosmos, the National Institute of Nuclear Physics (INFN) which used its own Stratasys Fortus 450mc FDM 3D printer to produce, at the National Laboratories of Frascati (Rome), the entire mechanical structure of the first cosmic telescope for UV rays. The telescope named "**Mini-EUSO**" (Multiwavelength Imaging New Instrument for the Extreme Universe Space Observatory) was launched on August 22, 2019, from the Baikonur Cosmodrome by the Soyuz MS14 spacecraft. We are now in the era of "everything is possible", two seemingly so different worlds like aerospace and 3D printers are now our present. Advanced 3D printing materials will further change the future of design and manufacturing.

As we have previously described, moments of "economic crisis" are also an opportunity to discover new horizons and new areas of development. And this is valid for all technologies including additive manufacturing. It is known to all of us scholars that innovation, while suffering from inertia, has generated antibodies to scenarios of uncertainty and unpredictable phenomena such as those of the lockdown, such as to be able to overcome them. If we had to associate an emblematic image of what was stated above, we would certainly choose the one shown in Figure 2.4. In fact, ventilators (which, as we all well remember, were unobtainable at that time) put the health system and especially intensive care in crisis. To deal with the emergency, some researchers, with a great spirit of inventiveness, *generated* by the innovation itself. In particular, they have created a solution of great importance in order to be able to save human lives in a world of emergency. Starting from a simple **Decathlon** Easybreath snorkeling mask, they managed to

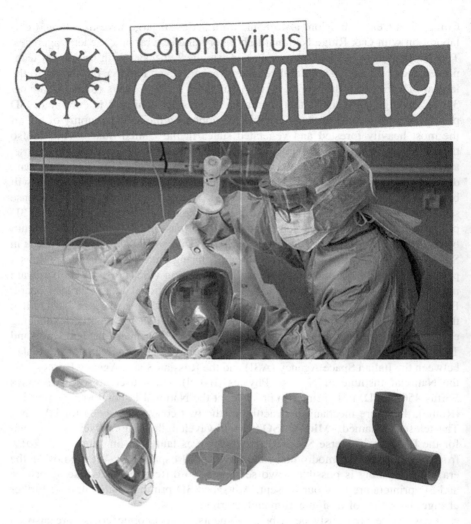

**FIGURE 2.4**   Charlotte valves for anti-Covid respirators.

create a hospital C-PAP (Continuous Positive Airway Pressure) device for sub-intensive care (https://youtu.be/w4Csqdxkrfw).

In our opinion, this is certainly the most fitting example of the capacity of innovation and of human minds predisposed to accept its capacity for contamination, to always find a way. However, it is a tortuous path to reach a possible solution. In the same period, the robotics researchers of the **University of Southern Denmark** (SDU, Denmark), given the virulent and contagious power of the coronavirus, created a robot (*swab robot*) *in additive printing* capable of performing throat swabs, so that operators and healthcare professionals were not exposed to the risk of infection.

3D printing has also emerged in recent years in the declination of **bioprinting** or **3D bioprinting**, i.e. as a technology for creating tissue and for regenerating tissue and organ injuries. Interestingly, traditional 3D-printed plastic objects are increasingly being used in biology. Today, bioprinting is making great strides and is set to be one of the most important revolutions in the global medical landscape. With 3D printing, it is possible to offer a customized solution by integrating a synthetic medical device with a bioprotected biodegradable one and also with autologous cells of the patient to improve regeneration. Currently, emerging innovations come from bioprinting cells or extracellular matrix deposited in a layer-by-layer 3D gel to produce the desired tissue or organ. Additionally, 3D bioprinting has begun to incorporate scaffold printing, a porous structure designed for cell deposition and proliferation. Scaffolds can be used to regenerate joints and ligaments. Ultimately, the advantages and potential are very interesting and range from **biomimicry** or the attempt to create functional body parts in laboratories to **mini-tissues**, to **self-assembly** which uses three-dimensional cell cultures to model the important functions of entire organs, with that of 3D printing. However, one of the current obstacles facing bioprinting is the vascularization of the printed tissue. The global potential remains enormous. We expect that in the near future, the medical field will be one of the main sectors for the development of 3D printers, which will obviously complement the traditional sectors.

## 2.4  SIMULATION AND CYBER-PHYSICAL SYSTEMS

The progressive digitization of all industrial processes, also due to the introduction of **Cyber Physical Systems (CPS)**, makes it possible to build a digital representation of the production plant that perfectly corresponds to the existing physical system or simulate a virtual system in part or in full. In other words, real and simulated data can provide vital information for decision-making processes. From this point of view, the **Digital Twins (DTs)** have established themselves and are establishing themselves as tools that have already been present in some way for many years in applied industrial research (think of the simulations of complex production plants). Today, DTs are greatly enhanced by the calculation of new processors and integration with new technologies.

The DT is a mathematical model that represents the system to be simulated and describes its key characteristics, behaviors, and functions of the physical or abstract system or process to be simulated. In this way, the DT makes it possible to test and understand the behavior of complex systems, products, or processes, even before realizing them or being able to modify the characteristic parameters of a process to be monitored or optimized. **Michael Grieves** first spoke about the digital twin in 2001, during a Product Lifecycle Management (PLM) course at the University of Michigan, in which he described the digital twin as the virtual and digital equivalent of a physical product. The **prerogative** of digital twins lies in the possibility of overcoming the difficulties encountered in a real laboratory, where the physical reproduction of the actual conditions of the system could require lengthy and costly process layout phases. Thus, the simulation becomes a sort of virtual laboratory, an experimental tool that is used in many scientific and technological fields to optimize times and costs. The possibilities for applications of solutions of this type are countless, but by way of example, in addition to the manufacturing sector, there are intelligent traffic control, cooperative robots, motoring, telecommunications, domotics, the functioning of health facilities, and the so-called smart grid. Demand for a simulation **software tool** that allows manufacturers to test new or redesigned production lines before flipping the "on" switch is soaring. Simulation technology is constantly evolving, and the next goals will probably include the development of software solutions integrated with AI algorithms in the field of robotics and process automation. The strong point of these tools is in fact that of adapting the offer to the technological demands of the market, with applications aimed, for example, at environmental impact assessments. Furthermore, it will become increasingly strategic **to support** the **customer in after-sales**. Today more than ever there is a need to simulate as far as possible not only the product, as was done in the past, but also the processes that pertain to production, which play a fundamental role.

## 2.5  INDUSTRIAL INTERNET OF THINGS

According to the definition given by **IBM,** the **Internet of Things** (IoT) is a neologism referring to the extension of the Internet to the world of concrete objects and places. IoT is a **new paradigm** that offers products such as cars, refrigerators, microwaves, thermostats, mobile devices, wearables, machines, animals, and people, the ability to be augmented with sensing and networking capabilities, enabling them to work together. The first concepts behind the IoT were outlined in 1982, when some researchers at **Carnegie Mellon University** in Pittsburgh, Pennsylvania, applied sensors and the network connection to a university soda machine to know its operating status. The term IoT was later coined by **Kevin Ashton**, who co-founded the Auto-ID Center at the Massachusetts Institute of Technology (MIT), where the standard system for RFID and other sensors was created. The IoT benefits from developments in the fields of electronics and wireless communication to enable communication and digital capabilities. Depending on the applications, the IoT may require the use of proximity devices, edge computing, for real-time data aggregation and processing. In addition, for interaction with cloud services, IoT is capable of performing sophisticated data processing, analysis on big data, machine learning, and AI. Ultimately, from utilities to healthcare, from production to public administration, there are now many sectors and working environments affected by the technological innovation of the IoT, with different levels of maturity. Among the sectors that benefit the most from IoT technologies are the automotive and transport sectors. Examples of IoT applications include remote control of the position and operating status of vehicles, protection of occupants in the event of an accident, insurance, and rental services. Wearable sensors have been made available to **De Beers Marine South Africa** (a division of De Beers, one of the world's leading diamond players) to monitor crew proximity to heavy machinery used in offshore diamond mining operations. On the ships that collect the precious stones in order to ensure the highest standards of operational safety for all personnel, De Beers has implemented the **Orange Business Services IoT solution**, developed to be able to achieve the ambitious "zero accidents" goal. In the industrial field, the use of IoT technology mainly concerns the development of products in the manufacturing sector.

The scope for expansion is wide. In the near future, connections to the IoT will require unprecedented levels of network connectivity. Therefore, standardization is required for interoperability and compatibility issues. However, a single standard will never be able to cover all cases due to the huge variety of applications. Alongside these problems, the work that is mostly being carried out on improving this technology is security, i.e. offering users certainty in the use and exchange of data.

Improvements in connection systems and the strengthening of the network have pushed companies more and more to find ways to exploit the "communication" capacity of objects to record information and data that could also improve industrial processes. In these cases, there has been talk of the **Industrial IoT**, even if today this terminology is this and has given way to the IoT alone. With IoT today we mean both the connection between objects that we commonly use in those three places where we spend a good part of our time (that is, home, work, and other places we frequent) considered strategic for sales by marketing and the dialog between machinery and equipment in companies. This strong extension of the concept linked to the possibility of objects communicating (with us and with each other) seemed evident to us as soon as we arrived at the first CES in Las Vegas. All the objects on display at CES were equipped with sensors capable of *communicating/transmitting information*. From **Rafael Nadal's** tennis racket which with some sensors on the cuff acquires so much information that it can quickly transform into a coach providing suggestions on how and where to best impact the ball, up to a fully integrated kitchen with virtual assistants such as Alexa or Google, able to communicate with the user to carry out operations at home, or independently decide whether to issue an order due to a lack of products at home or to cook a particular dish.

## 2.6   HORIZONTAL AND VERTICAL INTEGRATION

Complexity and instability are two elements that characterize the value chain. Therefore, change or the ability to adapt and innovation are two processes that guide today's companies in dealing with these critical issues. Among the various enabling elements and technologies that can well describe the digital revolution, there is the **horizontal integration** of production processes and **vertical integration** of production with other company areas. Innovating and promoting **horizontal and vertical integration** is a critical success factor that all companies should aim for. The goal should be to implement a transformation process that evolves static production systems into dynamic production systems ensuring a constant flow of data and information within and between companies and along the value creation chain. On the one hand, in fact, horizontal integration provides connected networks of cyber-physical and business systems that introduce new levels of automation, flexibility, and operational efficiency. On the other hand, through vertical integration it is possible to connect all logical levels within the organization, from after-sales, to production, up to R&D, quality control, product management, sales, and marketing. Horizontal and vertical integration is a natural consequence of technologies such as the IoT, cloud computing, and big data and has become so strategically important as to be considered the backbone on which to build a smart factory. In the idea of an integrated company, solutions supported by RTLS (Real Time Location System) marketing and monitoring technologies that allow for the automation of data acquisition play a decisive role. Some examples of these technologies are RFID, Beacons; Smart Labels; etc. These technologies can be applied, for example, in automated warehouses, in the use of robots or self-driving AGV vehicles.

With a view to integration and innovation, there is certainly a distinction between **B2B** and **B2C companies**. The ability to change very rapidly by creating direct contact with the market could represent a great opportunity for any industrial sector. However, the conditions must be created. Linking all logical levels within the organization is the foundation of a crucial competitive advantage in a global marketplace. At the same time, there is a need for a fundamental understanding of the activities and information flows within companies to enable seamless integration of networked production systems. From a technical and economic point of view, a digital integration is a key issue for realizing smart factories, as it allows all levels of a manufacturing enterprise to connect each other through a global information system with customers, suppliers, and other actors external. The potential for flexible integration is enormous. In contrast to existing inter-organizational systems, an organization must be able to meet individual customer-specific criteria in the design, configuration, ordering, planning, manufacturing, and operation stages. The goal is naturally ambitious and for this reason, requires appropriate interfaces for the integration of individual subsystems. This represents a significant problem because today many interfaces have a complex architecture and are difficult to use. Moreover it is still common practice that IT systems exchange information through extended interfaces, but can only use specific parts of this information. It is on the new opportunities deriving from the market, which focus on producing advanced technological solutions in the industry 4.0 field, with the aim of advancing within an increasingly competitive and cutting-edge market that aims to encourage and promote company production efficiency. Everything becomes simpler if we consider this type of integration as a natural consequence of other "4.0" elements, such as the IoT, cloud computing, and big data.

## 2.7 CLOUD COMPUTING

When PCs began to spread in the 80s, the underlying philosophy was to provide users with the power to process what they needed from a digital point of view directly from their desk. Subsequently, it was understood that the single PC could not perform all the operations in stand-alone mode. Thus began the experiments to create a computer network to provide a greater number of users with more important computing power and storage space. In the 1950s, with the "mainframe", users began to provide greater computing power and memory space. Initially, there was talk of timesharing, since those who wanted to use the performance of the mainframe for their calculations had to reserve computing time in which to complete their operations. Subsequently, compute instances with virtualization became abstract, purely virtual constructions. However, with the spread of the Internet, these virtualized environments were finally made available to all users. Only in the new millennium did businesses and individuals begin to familiarize themselves with the concept of the cloud. Google offered cloud services that were basically just file sharing and spreadsheet storage. **Amazon** subsequently began to make its huge server farms available to other users. Up to today, cloud computing has become part of the daily life of many people. Most smartphones are continuously connected to the cloud.

With **cloud computing**, purchasing IT services is cheaper and easier to manage and maintain than totally in-house solutions. Cloud computing can be defined as a set of technologies aimed at storing, processing, and transmitting data or, better still, the set of applications and software for dissemination to obtain useful information for the purposes of process and control of activity. It is therefore about storing and accessing data and programs on the Internet rather than on a computer's hard drive. The cloud is so defined because it can be imagined as an enormous cloud representing the gigantic server-farm infrastructure of the Internet capable of accepting connections and distributing information while it floats. Access to data is quick and simple: it can be done from any computer, mobile phone, or tablet wherever you are, as long as the device has an Internet connection. The substantial advantage that is driving more and more companies to adopt the cloud is found in management, in all its terms. It is no longer a need to purchase expensive software or hardware and no longer need large servers to store your data. In fact, the latter must be kept active 24 hours a day and this entails high management costs by the IT staff and, simply, electricity costs. By eliminating the servers, the company can have the necessary information available in a short time and almost everywhere and can improve productivity; the IT staff no longer involved in managing the company servers can thus be engaged in achieving more important and impactful business objectives strategic.

Cloud computing normally used for storage are **Microsoft OneDrive**, **Google Drive**, and **Apple iCloud**. Many examples of the use of cloud computing are also related to our daily activities. When we browse on Google by typing a search key, we are actually using cloud computing solutions. The computer does nothing but send the request to the network of one of thousands of Google computer clusters, which process the information *and ask our PC* to display the results. The most expensive operations are performed by a machine residing in California, Dublin, Tokyo, or one of the many other parts of the world without the user being aware of it. Even when we use e-mail via the Web we are browsing in the cloud. In the past, e-mails were consulted and sent through so-called clients running on the PC, which stored the messages on the disk space of your computer. With Gmail, Outlook.com, and many others, all this has changed and now online clients are used to consult, send, receive, and store all e-mail, saving it in some portion of server storage distributed throughout the world. Through this method, it is possible to consult e-mails anywhere in the world and in complete mobility. In principle, cloud computing involves **two components**: a cloud *infrastructure and software* applications. The former consists of the hardware resources required to support the cloud services provided and typically includes servers, storage, and networking components. The second component refers to software applications and computing power for running business applications, delivered via the Internet by third parties.

The potential and implications of cloud services are many and varied as demonstrated by the project developed by the Californian start-up **Cloud Constellation**. The idea was to develop a global data storage network located in space. **Space Belt** was created to ensure the security of its customers' data by storing them in a data center 650 km from the earth. Cloud Constellation believes that it can better counter cyberattacks in this way. It is a unique model. The project, which saw a joint venture between Leonardo and Thales, was the first data archiving and protection system in space created to evaluate potential requirements and cooperation scenarios in the field of operations relating to the ground segment of the SpaceBelt program. For the realization of the project, a constellation of ten low-orbit satellites was foreseen with the aim of protecting high-value databases of critical importance for highly sensitive activities, providing data archiving beyond the atmosphere, and managing network services with maximum security. By isolating its customers' data from inherently vulnerable terrestrial networks, the service promises the strongest mitigation of breach risk.

It should be noted that **cloud services** can be of different types depending on the needs. Without going into overly technical digressions, we can distinguish the services potentially provided into three types: **SaaS** (Software as a Service), **PaaS** (Platform as a Service), and **IaaS** (Infrastructure as a Service). One of the most popular models is SaaS (Software as a service). Through these configurations, the end user does not need any IT knowledge to use the application or the services provided. Classic examples of SaaS models that we use on a daily basis are Gmail, Google apps, and Office 365. IaaS and PaaS protocols are adopted by companies such as Netflix, Spotify and Airbnb which have exploited these systems to face the exponential growth of the business. In the first case, i.e. of IaaS models, the provider offers virtual hardware (CPU, RAM, space, and network cards) and therefore the flexibility of physical infrastructure, without the burden on the user of hardware management. This type is dedicated to system administrators or systems engineers. In the second case, ie that of PaaS, the service provider takes care of the hardware infrastructure, while the user will have to install the operating system and develop his application.

## 2.8   CYBER SECURITY

Providing a definition of **cyber security** is not easy. Let's try to explain the term through some examples. Let's imagine ourselves in the guise of a Uruguayan structural designer. Uruguay is a country recognized by seismologists as an area with very low seismic risk. Therefore, this land can be considered anti-seismic to all effects. The regulatory references and design rules are not the same as those used and applied in countries such as Italy or even worse in Japan, where seismic events are the order of the day. Therefore, if we built a building in Japan, following the design principles used in Uruguay, probably, at the first earthquake shock, this building would collapse within seconds. This example helps us understand what happened in 2016 to Fiat Chrysler Automobile (FCA), now Stellantis, which was forced to "recall" over one million cars and SUVs distributed in different parts of the planet, following an inspection carried out by the same manufacturer, and after several reports from users, a manufacturing defect had been found. In fact, some models had a problem in the automatic gearbox control device which signaled the car to be in "parking" mode even when in reality the command was not engaged with the consequent risk of leaving the car without the brakes once parked.

**What do these two examples have in common?** Well, wanting to simplify, we could provide an answer to this question by following these logical steps, starting from the beginning. Let's start from the origin of the introduction of technology and sensors in the automotive world. One of the first sophisticated experiments in installing sensors on a car was certainly the introduction of an electronic control unit to control and monitor the operation of the ABS together with other functional control devices related to adaptive cruise control. Once the potential of these sensors was recognized, car designers successively proceeded to connect various car devices to the control units, increasing the interrelationship between the various systems inside the car so as to provide as much information as possible to the driver. The problem arises when some brilliant engineer, rightly from his point of view, begins to hypothesize and then realize the possibility of connecting the car with mobile devices, such as smart phones, for example. It is here that a serious "design" error is made from the point of view of IT security since the "closed world of the car" opens up to the outside. By itself this condition, as seen by an auto engineer, presents no problems. indeed only an opportunity to further enhance the customer's customer experience. In reality, we find ourselves in the same condition as a possible imprudent Uruguayan engineer engaged in designing the building in Japan according to the design schemes of his country of origin. In the same way, the car designer, if he is not careful in creating an information security infrastructure, obtains the same nefarious results, since he risks opening a huge door to some malicious person, who could thus have access to the car through this portal.

And hence the need for FCA to withdraw several of its models since they had suffered a computer attack that had led to the problems described above, and fearing that things could get even worse, the manufacturer deemed it necessary to implement a recall campaign. In other words, today's cars are highly technologically complex systems that allow the car to always remain connected and communicate in real time with the surrounding environment. However, these highly networked systems have many vulnerabilities because they are not closed systems, and therefore allow access to the vehicle from the outside.

If, on the one hand, remote control and sensor solutions of all kinds are experiencing exponential growth, it is equally evident that the use of connected devices can paradoxically represent a risk from the point of view of security and not just information. With the advent of 5G, autonomous driving will be widespread. Imagine what could happen if the autonomous driving protocol were violated. Someone could control our cars remotely regardless of our will. If we think about it, this is what could also happen to air traffic control today, without wanting to be too fatalistic. In fact, if we remember after the attacks of September 11, 2001, a system was introduced in every aircraft that prevented anyone from gaining access to the cockpit. The new systems installed in each aircraft allowed (and allow) the opening of the same only through a command activated from inside the cockpit, preventing access from the outside. However, after a few years, in 2015 **Andreas Lubitz**, first officer in force at Lufthansa suffering from depressive problems, used this "safety" device to barricade himself inside the cockpit, piloting the aircraft until it crashed on the impervious mountains of the locality French Prads-Haute-Bléone, at an altitude of about 1,500 m with 150 passengers on board. It was Germanwings flight 9525, an international scheduled flight of the German low-cost airline Germanwings, in service between Barcelona (Spain) and Düsseldorf (Germany). Well, the solution found to this new problem by the engineers (also guided by the approach of the naive Uruguayan engineer) was to be able to activate a remote release system so that the control tower could take possession of the aircraft controls at a distance and take him to a safe place. Meritorious objective, of course, but, from a cybersecurity perspective, we immediately realize the threat hidden behind such a solution.

We can affirm, without a shadow of contradiction, that cybersecurity in our current world, increasingly digital and connected, is an important issue not only reserved for insiders or large companies but is also a social issue.

**At this point we must clearly define what is meant by cybersecurity.**

We can define cybersecurity as the set of tools, skills, and practices that allow an organization to erect a barrier between the outside world and the information it possesses in digital format. Our organizations are increasingly digitizing their processes and their services and the entire ecosystem connected to their production: PCs, smartphones, databases, platforms for productivity and communication, but also tools dedicated to the front-end, such as e-commerce portals and telephone switchboards, as well as servers for the management of much more sensitive data related to finance or justice. The theft or blocking of a dataset necessary for the performance of a strategic task or a hacker attack (to be distinguished according to the targets and purposes as we will see later) capable of paralyzing the systems can cause operational damage, which will inevitably on business results. An important change, compared to what has been recorded in recent years, is that intrusion threats are no longer reserved only for large-sized organizations, or more structured on the IT front. In the crosshairs of hackers, or at risk of disservices due to internal errors, today we are all both SMEs and each of us, as long as we have an Internet connection or a smart phone. The latest CLUSIT Report (a document produced by the Italian Information Security Association since the 2000s) on **ICT security,** report how in Italy cyberattacks are increasing year by year without interruption. This Report, now in its tenth year of publication, provides us with a significant overview of the most significant cybercrime events that have occurred globally in recent years, comparing them with the data collected in the previous 5 years. The study is based on a sample made up of more than 10,000 known attacks of particular seriousness (of which 1,670 only in the last year, i.e. that have had a significant impact on the "victims" in terms of economic losses, to reputation, of disclosure of sensitive data (personal and otherwise), or which in any case prefigure particularly worrying scenarios, which have occurred in the world (including Italy) since January 1, 2011. Compared to 2018, the trend is continuously growing, and in absolute terms over the past year – the highest number of serious attacks is observed toward the categories "Multiple Targets" (+29.9%), "Online Services/Cloud" (+91.5%), and "Healthcare" (+17.0%), followed by "GDO/Retail" (+28.2%), "Other" (+76.7%), "Telco" (+54.5%), and "Security Industry" (+325%). The study also highlights how the FINANCE sector, in Europe, continues to be one of those preferred by cybercrime. It is interesting to record how the attack techniques see a significant diversification. The *malware technique*[3] for harming organizations remains in the lead, used in 44% of attacks, with constant growth over the years. However, it is interesting to note that there are still unknown techniques, as well as social engineering and phishing, which are also showing continuous growth trends. Finally, they have a mention in the various reports following both the exploitation of vulnerabilities and APT (Advanced Persistent Threat) attacks. *Ransomware*[4] accounts for 46% of malware attacks, while *cryptomines*[5] are down. Individual organizations are affected, but also millions of people simultaneously, with a more significant increase in generalized attacks compared to targeted ones. According to CLUSIT experts, ransomware has prompted entire states, such as Louisiana, to even ask for the declaration of a national emergency. To understand the impact of these techniques, severity analysis is helpful. The **CLUSIT report** divides attacks into three tiers to assess the severity of incidents. It emerges that attacks with medium impact are the majority, with 46% of cases. The most serious represent 28% of cases, and those of critical level represent 26%. Among the targets, critical infrastructures are those that have mainly suffered attacks of a critical level. Finally, it is interesting to record how the dawn of this new decade saw the "birth" of **deepfakes.** Deepfake is an AI-based technique for human image synthesis, used to combine and overlay existing images and videos with original videos or images, via a machine learning technique, known as a generative adversarial **network**. Deepfakes have been used to artificially impersonate famous politicians on video portals or chatrooms. For example, Argentine President **Mauricio Macri**'s face was replaced by **Adolf Hitler's, and Angela Merkel's** face was replaced by **Donald Trump's**. In April 2018, Jordan Peele and Jonah Peretti created a deepfake using **Barack Obama** as an advertisement about the dangers of deepfakes (Figure 2.5).

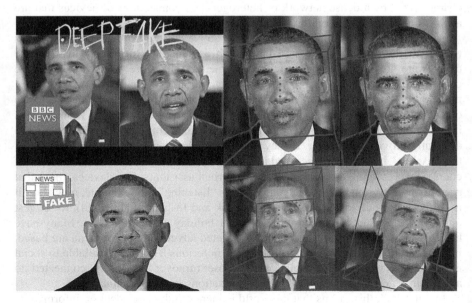

**FIGURE 2.5**   Deepfake Obama.

Mark **Zuckerberg** circulated on the net is also famous. The video, created by Bill Posters and Daniel Howe in collaboration with the advertising company Canny, shows Mark Zuckerberg sitting at a desk giving an eerie monologue about the power of Facebook. Superimposed on the video, there is an inscription that replicates the style of an American newscast, which reads *We are improving transparency on advertising*, as if the whole were actually part of a news service.

According to the report CLUSIT, the deepfake represents one of the main cyberthreats of the next decade. The reason why deepfakes constitute a phenomenon that is under consideration can already be traced to the very term that represents them: the "deepfake" is a deep fake, a set of video, audio, or audio-visual contents counterfeited in such a sophisticated way as to touch perfection. The main problem that deepfakes embody is that with them we end up seeing the ability to judge almost completely crumbled, i.e. the autonomy to distinguish between what is authentic and what is not. Analysts agree that deepfake attacks on business organizations will be inevitable. However, there are some measures that companies can take to combat the phenomenon. First, company employees must be educated. In our view, the biggest challenge in the field of fraud no matter where it comes from is regulation and the creation of new regulatory and technological standards aimed at harm limitation. However, the concerns that they could be an effective tool for scams or manipulations certainly do not escape anyone. According to the report, a phase shift has taken place in global levels of cyber security, caused by the rapid evolution of the actors involved, the methods, and the purpose of the attacks. In fact, the integration and connection of devices is characterized by IoT vulnerabilities, also due to the recent evolution of 5G technology and networks. The IoT is the increasingly widespread and consolidated representation of a digital ecosystem characterized by a dense network of heterogeneous connections of devices that are always connected and therefore potentially vulnerable.

A very pertinent example that allows the above concept to be expressed well is related to the use of virtual assistants such as **Google Home** or **Alexa**. These devices are now present in almost every home, and therefore, even you, will have occasionally experienced an involuntary activation of Google/Alexa. This phenomenon is linked to the fact that these devices are continuously connected to the outside world and continuously acquire test data of vocal information. If among this information they record, they recognize the predefined command they take action to diligently provide an answer. If they don't recognize the command, they get rid of this "useless" information, recorded in a memory that we could define as "short". Of course, this reasoning is valid not only for virtual assistants. Our devices, those that accompany us throughout the day, are always "listening". For example, nowadays it is common for our social channels (Facebook and Instagram above all), to suggest our favorite products (food, clothes, ...), our holiday destinations, etc. Many have had this same experience. Most of those targeted advertising phenomena are based, in fact, on the enormous amount of data (the notorious Big Data) available to social networks, as well as other platforms whose turnover is closely connected to advertising, and for which our futile conversations can have value.

In a nutshell, this means that the world is increasingly dependent on information, and digital is increasingly dominating our planet. Currently, globally, we produce 10 to 21 bits (10 followed by 21 zeros) every year. If we assume a growth that follows Moore's law, in a few years the number of bits produced will be more numerous than the atoms of the Earth, and it is estimated that the energy needed to keep them in "motion" will be equivalent to our current total consumption of energy. In 2070 there will be at least 1 kg of bits stored in some cloud for each of our personal devices such as PCs, cell phones, or tablets. According to Melvin Vopson, at the School of Mathematics and Physics of the University of Portsmouth (UK), the bit is not an abstract entity but something physically concrete, following Einstein's principle according to which mass and energy are equivalent. According to the English scholar, all of this could lead us to consider "information" a fifth state of matter together with gas, liquid, solid, and plasma. And this, infoabulimia, could generate a further catastrophe, which he defines as an "information catastrophe", which will complement the climatic, energy, and population movement catastrophes.

In reality, this "catastrophic" aspect was already highlighted several years ago. As often happens, fiction anticipates reality (as we have also seen with robotics or the conquest of space by way of example only), and also in this case, 20 years ago, in 1992, in a legendary film, "The Fraudsters", with **Robert Redford** then as Martin Bishop at the head of a *penetration test* (or informally **pen test**) agency specializing in analyzing and evaluating the security of a computer system, the same states that the world is no longer dominated by weapons, from energy from money, but from many 0's and 1's, from bits. Furthermore, he adds, that in the ongoing world war, "it will not be whoever has the most bullets who wins, but those who *control the information*". Large multinationals are also the ones that have suffered the most attacks, which have also led to serious economic losses. This is precisely the point, in order to counter these phenomena and those who govern them, first of all, we must understand who we are fighting against, as **Sun Tzu said** in the famous book *The Art of War*, the first step to victory is to know your enemy. Unfortunately, we don't know about it. And often from the press and television, we have learned of a "model enemy" who is a far cry from the classic nerdy, adolescent, male, with relationship problems, and even with a black hoodie. If we want to base our defense, imagining that we have to defend ourselves against him, then we are on the wrong target. According to several experts, we can associate the hacker that we recall in our minds with the **industrial spy**, i.e. very competent and who "attacks" someone because he has a specific interest. However, the greatest concern must come from **organized crime** which has generated a new criminal model, cybercrime. They all attack horizontally and wait for someone to "bite" at one of the millions of phishing messages that arrive in our e-mails. Cybercrime is aware of the fact that statistically 8% (still today) of the people reached click on one of the links forwarded to us, and by entering their credentials on the net, these people can generate a significant profit for the organization. Thus, the dimensions of the phenomenon are enormous. In the CLUSIT report, what was said is summarized with a sentence that we quote below: *it doesn't matter who you are, it doesn't matter what you do, it doesn't matter what you do, they will attack you.* We worry about defending ourselves "from nerd hackers", and we don't pay attention to cybercrime. Its growth leads us to think that the costs of e-mails are becoming such that it could be cheaper to send a "letter" by traditional post the old way than sending an e-mail. We must be aware that this type of crime is not improvised but evolves and improves.

Think that there are people who offer money, for example, to an Apple employee to provide his username and password (reward set at $23,000). This means that there are criminals who have evaluated the investment, established its size, and have also evaluated the entrepreneurial risk associated with the economic outlay. Evidently, they have a clear understanding of the revenue model and the ability to return on the investment generated by the possession of information of this nature. We often have no idea what damage a cyberattack can cause. Computer scientists, until a few years ago, didn't even have an expression to size the gravity of the event of an information attack. They borrowed the term from the world of economics, and today even IT experts define catastrophic events as extremely rare, but which, if they occur, bring the companies that suffer them to their knees, with the name of Black Swan. A recent example of a "black swan" was that of EQUIFAX, one of America's largest consumer credit check companies (credit check). Equifax collects a series of financial information on consumers and calculates their "credit score", a score of reliability, establishing their creditworthiness with respect to any credit requests. In 2017, the US company suffered a cyber breach that put the data of 143 million American citizens at risk (in addition to a certain number of data on Canadians and Britons). The data subject to attack ranged from simple customer records to driver's licenses and credit cards. In addition to involving the theft of hundreds of thousands of customers' data, this intrusion caused a loss on the stock market of 35% of the value of its shares in just one week.

But the expertise of cybercrime goes much further. If the stock market can be affected by attacks suffered by companies, then why not *simulate* such attacks and organize yourself to earn on stock market fluctuations?

Thus, in April 2013, someone stole the **social express password**, i.e. ANSA of the United States, and relaunched a news story relating to an explosion in the White House of two bombs, reporting that the President had been injured during the explosion of the United States. A real bag manipulation. This is just one example. Unfortunately, there are several cases of this kind at national and global levels. We must therefore understand that if we are in control of the process, we are the ones who can defuse attacks, or at least limit the damage.

According to the CINI National Cybersecurity Laboratory, cybersecurity, according to the former president of the European Commission, **Jean-Claude Junker**, is the **second emergency** in Europe, after climate change and before immigration. In reality, for several years chancelleries around the world have put cybersecurity at the very top of their agendas (even the last B20 meeting in Saudi Arabia saw cybersecurity among the main topics reported in the recommendations produced by the task forces). Blocking the operations of companies, surreptitious control of critical infrastructure services, andtheft of intellectual property or information crucial to the survival of a company are examples of the major threats a country faces. Recent *malware attacks* **WannaCry** and **NotPetya** have been the visible events of an impressive series of attacks in every corner of the planet.

Cyberspace is the most complex thing that man has ever built: on the one hand, the union of thousands of networks that make it difficult even to have a snapshot of whoever is connected to them; on the other, the stratification of software programs and protocols developed in the last 40 years. This complexity generates vulnerabilities (software errors, misconfiguration, and protocol weaknesses) that are exploited by cybercriminals to steal data or cause damage. In an increasingly digitized world, cyberattacks raise alarm among the population, cause enormous damage to the economy, and endanger the very safety of citizens when they affect distribution networks of essential services such as health, energy, and transport, namely, the critical infrastructures of modern society. Imagine what could happen if all the traffic lights in a metropolis suddenly went out, the elevators stopped, and the ambulances could no longer receive the right address to recover the injured. Many times the damages of cyberattacks depend on an identifiable weak link. The weak link in cybersecurity is the **human factor**. Man is now an integral part of cyberspace, and therefore the human factor represents the most important and unpredictable vulnerability of this macrosystem. A wrong click can in fact destroy any line of technological defense of a single device, an organization, or a country. These are the people who get "fished" by a phishing campaign, who use the name of the cat or their spouse as a password, who use the same smartphone to let their children play with it and then access the company network. These people are the first to open the doors for criminals to their organization's sites, networks, and databases, with dangerous and unpredictable effects. Before the advent of cyberspace, the world was based on information printed on paper or stored on isolated computers located in well-defined physical perimeters. This world has developed very precise threat models, allowing the definition of sufficiently clear and detailed national, corporate, and individual safety and security policies. In cyberspace, however, threats are constantly changing, and many remain unknown for months or years before emerging. We therefore find ourselves having to define security policies in a world where information on the threat is highly incomplete. A country that does not place cybersecurity at the heart of its digital transformation policies is a country that puts its economic prosperity and independence at serious risk.

## 2.9  BIG DATA AND ANALYTICS

It is particularly interesting to explore the reality of **Big Data** as a tool and as an essential resource for cultural and scientific progress and evolution. Man has always had an interest in keeping information in order to be able to consult it later, as confirmed by the creation of the Abacus, the library of Alexandria in Babylon, as well as many other historical episodes. In **1965, the first data center** was created in the United States, and only a few years later a developer created the first framework for a relational database, i.e. a "warehouse" of data where the various tables containing the data are linked together through reading keys. Digitization is significantly and exceptionally increasing the possibility of analyzing and collecting data. It is an exponential phenomenon. We are in a unique historical moment. There are interesting examples of how big data is able to increase efficiency and create a surplus for consumers. In fact, the computerization and digitalization processes (from a smart factory perspective) have always represented a decisive value in the choices of manufacturing companies.

Personally, we began to have a perception of the big data phenomenon when we started connecting to the web in the 90s. At the time, the use of the web was essentially linked to our academic research activities. The amount of data and information that could be collected was much greater than the possibility of consulting books or paper magazines. Indeed, it was precisely at the end of the 20th century, more precisely in 1991, that we witnessed the birth **of the Internet**. It is an epochal change for the first time the web offers the possibility of making data accessible to everyone and everywhere in the world, and thanks to technological developments, digital becomes cheaper than paper for the first time. The accessibility of information concerning any subject, even specialized subjects, previously limited to insiders, is now "ubiquitous" and at no cost. It was precisely the 90s that established themselves as the determining years in this matter, consolidating this growth in 1999, the year in which **the word "big data" was read for the first time** and people began to talk **about the IoT** and the possibility to connect objects connected to the Internet to each other. Then only a few visionaries would have imagined the vast use we are making of it in 2020 and that we will most likely continue to make it in the future. The day in which Italy entered for the first time in the "network" that would change our lives forever, dates back to April 30, 1986, thanks to a group of researchers from Pisa engaged in a CNR project, the National Research Center of Electronic Computing. If at the time we had imagined how much the Internet would have revolutionized our lives, perhaps the news would have caused more sensation, but in truth hardly anyone noticed. That day, the University of Pisa got in touch with the satellite station of **Roaring Creek**, in Pennsylvania, thanks to the **"Butterfly Gateway"**, a computer with the name of a butterfly but the size of a wardrobe. In any case, we can claim the fact that we were among the first countries in Europe (precisely the fourth) to access the network. In the last decade, the greatest transformation of the Internet has been the possibility of **surfing from mobile,** wherever we are.

The great potential of big data is not only to accumulate large quantities of data which, taken together, occupy a lot of storage space in the order of terabytes, but above all to be able to process them. Without the ability to process data, they wouldn't have much use. Think, for example, of analyses on **national census** data or real-time analysis if you want to obtain the trend of any phenomenon related to voting preferences during political elections or the trend of data during a pandemic period (such as, for example, the one lived in 2020). Increases in data collection and growth in processing power complement each other. One does not prescind from the other. According to the analyst **Douglas Laney,** when it comes to big data it becomes impossible not to mention the famous "3Vs", defined for the first time in one of his studies in 2001. Over the years, two other elements have been added, which define the "5Vs" which today most represent this context: volume, speed, variety, value, and truthfulness (Figure 2.6).

When we talk about data, however, we must distinguish between *structured data* and *unstructured data*. In order to find a way to extract value from unstructured data, it becomes necessary to be able to identify which actions can prove to be decisive and which are useless, skimming as much as possible a quantity of data so numerous as to be quite difficult to decipher.

## The characteristics of Big Data

| **VOLUME** | **SPEED** | **VARIETY** | **TRUTHFULNESS** | **VARIABILITY'** |
|---|---|---|---|---|
| HIGH VOLUME OF DATA (>50TB) | DATA GENERATED AND ACQUIRED QUICKLY | HETEROGENEOUS DATA BY SOURCE AND FORMAT | DATA QUALITY AND RELIABILITY | CHANGEABILITY OF MEANING DEPENDING ON THE CONTEXT |

**FIGURE 2.6**   The 5Vs that characterize Big Data.

**But what is the real function of using big data?** In our opinion, the answer that best summarizes the question is to provide the best possible representation of reality through data. **Examples of big data are many in practice.** With big data it is possible **to collect information that was unimaginable until a few years ago**, such as reviews, comments on social networks, or user behavior data on the company website, which makes it possible to profile the customer on the basis of the attitude with which he addresses the brand and no longer using only static variables (typically demographic). A typical case is **location-based marketing**, i.e. the use of geo-location data from smartphones and other mobile devices, to identify the customer's position and then carry out real-time advertising actions (an example is reported in the previous chapter). In marketing, the use of big data is familiar in the construction of so-called **recommendation methods**, such as those used by **Netflix** and **Amazon** to make purchase **proposals** based on the interests of a customer compared to those of millions of others. All the data coming from a user's navigation, from his previous purchases, and from evaluated or researched products allow the giants of commerce (electronic and otherwise) to suggest the most suitable products for the customer's purposes, those that tickle his curiosity and push him to buy for momentary, permanent, or simple impulse needs. A final example, perhaps the most fitting, is the project conducted by Google on the new campus in Mountain View, California. By analyzing the groups of search terms entered by users on its engine, he was able to predict (only in 2008) the progress of influenza outbreaks in the US territories faster than the Ministry of Health itself was able to do using the hospital admission records of public and private health facilities. Today, on the basis of Google's experiment, GSK, one of the most important pharmaceutical companies in the world, has used big data (a set of data from individual conversations, forecasts, and statistics from recent years, access to pharmacies, requests for information, or contacts with doctors) to create an algorithm that would make it possible to anticipate the outbreaks of seasonal flu in Mexico so as to concentrate distribution and advertising exactly in the areas identified by big data readings. This all sounds so sci-fi, but, in reality, it's only a tiny fraction of what big data could do. However, one must also be aware that data processing algorithms are not always that precise. Suffice it to say that in 2014 big data failed to predict the Ebola epidemic. The reason is essentially linked to the fact that the data arriving from the areas of West Africa affected by the virus were presented in languages that the control programs did not collect. In any case, the multiple uses of big data lends itself to innumerable applications in the field of health. The potential of big data is enormous, but when their use is governed by human judgment and human logic, such data are by nature devoid of soul and creativity. A study conducted by a group of researchers from the American University of **Harvard** had already hypothesized – by monitoring Internet searches and analyzing the flow of accesses to the emergency room – that in Wuhan, China, COVID-19 had already been present since August of 2019. The main obstacle to overcome is the distrust of companies, research centers, and some scientists to share the data on which big data could work. There are many examples of the use of advanced data analysis to reduce the costs of production processes in manufacturing contexts. First of all, by constantly monitoring the data coming from the sensors of a digitized factory, it is possible to understand the state of health of all the machinery and therefore predict its breakdown; this is predictive maintenance planning. Data in the past was mostly tables or databases generated by corporate information systems. The data of the present are many more and take different forms, not only well-defined rows and columns but also texts, images, videos, log data, geolocation data, and so on. The data of the future – and in some cases already today – will become the main enabling element of a huge technological revolution. An example of an application that is attracting great interest in the banking and insurance world concerns the identification of frauds. The case of **ZhongAn is very interesting**, a company founded in 2013, the first Chinese company known to us to have obtained the license to operate as an exclusively online insurance company in China. ZhongAn has no physical offices and conducts all activities online (mostly on mobile), including litigation. But what is more interesting is the automation and sophistication of the back-office, totally based on the exploitation of big-data and blockchain databases. The system designs the policies and premiums on the basis of a continuous analysis of the market, the risk, the applicant's history, and many other inputs (all data automatically obtained from millions of digital pieces of information). Not surprisingly, the founders of the company are Alibaba and Tencent (WeChat) and PingAn (the largest Chinese insurance company). ZhongAn already has over 400 million customers, and it certainly has no plans to stop its growth.

An interesting example of **the combination of big data and artificial intelligence** is offered by **Nick Brestoff**, founder of **Intraspexion**, a newly conceived company in the legal sector that seeks to predict and prevent potential disputes before they happen. Intraspexion is, in other words, an *early warning system* that helps companies avoid litigation. The system works through a deep learning algorithm trained with several mails and sentences of generic lawsuits so that the same software can predict the risk of filing a lawsuit and avoid it. A system that recalls **Steven Spielberg's** 2002 film, *Minority Report*. However, in the film, the police relied on the premonitions of three individuals with amplified extrasensory powers of precognition, the famous **Precogs**, to prevent the murders and to arrest the potential "culprits". Now Intraspexion uses sophisticated algorithms and a huge amount of data instead of the famous Precogs. This is the aspect that makes you think and which contains enormous potential. Thus we go from fiction to reality.

We can state that the usefulness of completely disaggregated data from each connected object would be completely null if not combined with analysis algorithms (Data Analytics) combined with artificial intelligence systems. This combination gives rise to different solutions in every field: from food to gamification, from industrial production to advertising, and from technical to humanistic professions. With particular reference to "humanistic" professions, online instantaneous translators have certainly begun to revolutionize a part of these professions. **Google Translate** is one of the applications in continuous improvement given the persistent and uninterrupted use that is being made of it and the ability of the algorithm to improve continuously. Google Translate turned 11 on April 28, 2020, but it seems like we've always known this interpreter who can translate in just one click, articulated and complex sentences in over 100 languages, from Afrikaans to Zulu. Over time, Google's translation system **has evolved enormously due to its use by over 500 million people, with more than 100 billion words a day and over 3.5 million continuous users, which by now have constituted a real community.** The use of this data has allowed and will allow Google to improve its translations more and more and therefore its usefulness also on a professional level.

## 2.10 ADVANCED MANUFACTURING

With the term **Advanced Manufacturing Solutions** we mean all those advanced production systems, interconnected and modular, which allow flexibility and performance. These technologies include automatic material handling systems and **advanced robotics**, i.e. the inclusion of collaborative robots on production lines within the logistics-production process, to make them more efficient and competitive. Thus, "advanced manufacturing" is a term used to describe manufacturing processes that are based on cutting-edge discipline and technological research. AM includes the development of manufacturing techniques and specific new technologies. This results in many advantages, including the minimization of the resources used or the removal of project constraints, a reduction in energy consumption, and, above all, in toxic and harmful substances for the environment. Adopting Advanced Manufacturing Solutions also means offering the possibility to exploit a number of emerging business models, which can positively alter the structure of the manufacturing sector. SMEs constitute the backbone of our production reality and technological innovation and therefore digitization in all its aspects, represents an obligatory path for their survival in the first instance and then for their growth. Precisely for this reason it is essential to create the conditions so that, even in times of crisis, SMEs can have the ability, autonomy, and strength to support their structure in a depressed market. Collaboration with research centers and technology transfer centers could represent a great opportunity to make SMEs more competitive internationally and support them in their growth. From this point of view, even the level of automation (for example the introduction of collaborative robots) must be strongly weighted on the basis of the actual benefits that can derive both from the point of view of productivity as a whole and from the point of view of the positive impact on human resources.

Given the vastness of the topic and its connection with collaborative robotics ("cobots" interconnected and programmable), we postpone the deepening of this world in a subsequent chapter dedicated to automation and omination.

## NOTES

1 Ivan Edward Sutherland (Hastings, May 16, 1938) is an American scientist and computer scientist, Internet pioneer, and winner of the Turing Prize in 1988 for the invention of software and Sketchpad, predecessor of the interfaces most used in computer graphics. He is also responsible for the design of VR glasses.

2 Chuck Hull (born Charles W. Hull) (Clifton, May 12, 1939) is an American engineer, inventor, and entrepreneur; he is the co-founder, vice president, and chief technology officer of 3D Systems.

3 Malware is a term coined in 1990, an abbreviation of the English malicious software, lett. "malicious software", usually translated as malicious software. In computer security, it indicates any computer program used to disturb the operations performed by a computer user.

4 Ransomware is a type of malware that restricts access to the device it infects by demanding a ransom to be paid to remove the restriction.

5 Malicious cryptominers belong to the category of malicious code designed to hack the idle computing power of victims' devices and use it to mine cryptocurrencies. Victims are not asked to consent to these activities and may not even be aware of what is happening in the background.

# 3 Emerging Technologies

## 3.1 ARTIFICIAL INTELLIGENCE

One of the most influential and important scientists in artificial intelligence (AI) is **Kai-Fu Lee**[1] who worked at Apple (from 1990 to 1996) developing the first **voice recognition system** 25 years before **SIRI**. In his most famous book *AI Superpowers: China, Silicon Valley, and the New World Order*, Lee explains with startling clarity how in less than 20 years AI has emerged as the powerful force it is today and how China is rapidly becoming the world leader in AI. It is good to specify, after this premise, that the interest in **AI** originates with the birth of the first calculators, but the foundations of the concept of AI were born in **1936**, the year in which the father of modern information technology, **Alan Turing**,[2] formulated the hypothesis of a machine capable of carrying out any type of calculation. The scientist was the first to question *the existence of the faculty of thought in machines*. The term AI was coined only later by **John McCarthy**[3] in 1956. McCarthy claimed that within 10 years a machine would be able to prove mathematical theorems or beat a chess champion and explain human behavior. It actually took 50 years to achieve just one of the goals. In 1996, IBM's **Deep Blue** was the first computer to win a game of chess against reigning World Champion Garry Kasparov. One of the best-known computers in the world today is **Watson,** the successor of Deep Blue. It is an AI system capable of answering questions expressed in a natural language. Watson has risen to the headlines for having defeated the world champions of the **game, Jeopardy**. It is a US television quiz show, still much loved by Americans today, which consists of a general knowledge competition between the various competitors, broadcast on NBC since 1964. Watson, sitting in the seat of a competitor, was able to answer exactly the general knowledge questions of the TV program, up to winning 1 million dollars, which was then donated to charity.

The game is divided into three phases, and in each phase clues are proposed to which the participants must offer their answers. But the great difficulty is that the answer must be given in interrogative form, or it will be judged incorrect. Therefore, we understand how difficult it can be for a computer to *understand the meaning of the sentence* to be able to provide the answer in the form of a question.

Today Watson is also used in particular ML processes in healthcare. The IBM super computer is able to analyze x-rays and make diagnoses, until it identifies a possible prophysalli. Watson also transformed into a foodtruck, the orange pickup truck parked in Austin, Texas, trying recipes following the advice and input received from users via Twitter. But perhaps one of the most interesting applications is the **Ross project**, born in collaboration with the University of Toronto, which has applied Watson's super-intelligence and its ability to analyze large amounts of data to solve and study legal cases. Fed on Ontario data and laws by university students, Watson-Ross was challenged and studied interesting cases.

DOI: 10.1201/b22968-3

Today it is an *app* available to lawyers. Thus the American law firm Baker & Hostetler "hired" Ross to manage bankruptcy law practices. Ross is, in fact, able to understand the questions and give an answer by quoting legal rules and sentences (University of Toronto – Watson University Competition Demo, https://youtu.be/ ODPgh4Jlv_I). Watson is also able to analyze tweet data from all over the world, making his brain available to companies that want to study trends and thus wisely direct their business. One of Watson's greatest challenges was learning a foreign language very far from English: the computer memorized Japanese terms and grammar and was educated to think in that language.

### 3.1.1   AI: WEAK AND STRONG

Automation, increasingly articulated, has given rise to two fundamental theories: **weak** AI and **strong** AI, so-called thanks to the American philosopher **John Searle**. The distinction between strong AI and weak AI starts from a fundamental question: *will the machine be able to equal and even surpass human reasoning or will it never be equivalent to it?*

Weak AI is based on the "as if" or acts and thinks as if it had a brain. The goal of weak AI is not to build machines that have human intelligence, but rather systems that can perform successfully in some complex human functions. While, in strong AI, the machine is not just a tool. If properly programmed, it itself becomes a mind, with a cognitive capacity indistinguishable from the human one. The technology behind strong AI is that of expert systems. Until the late 1980s, scholars believed strongly in strong AI. An example above all, **Isaac Asimov's** famous robots. The scholars who most believe in this branch of AI argue that it is possible to create an artificial machine that is, in all respects, equal to and superior to the human mind. However, we ask ourselves *until* a *machine has* a complete understanding of the meaning of natural language, can a machine ever be called intelligent? We are certain that we will see great changes in this area and that the development of technology will project us into scenarios that are unimaginable today. Suffice it to say that as far as web research is concerned, Google relies on a self-learning component of the algorithm for interpreting user requests. Since 2016, the entire Google platform has been integrated with the **RankBrain algorithm**, an AI system capable of "learning" independently. The main task of RankBrain is to interpret keywords and phrases with the aim of identifying the respective user intention. **Facebook,** on the other hand, uses algorithms to try to process the data provided by our searches and offer us a wide range of news and ideas that can stimulate our attention. **Profiling systems** are another example of the evolution of the algorithms underlying AI. Through the application of algorithms, the Internet knows how to develop and provide us with a model capable of suggesting, based on our interests, the music to listen to, the books to buy, and so on.

### 3.1.2   HOW AI WORKS

But how does AI work, and what do machine learning (ML) and deep learning consist of? Actually, to understand ML and deep learning we should first define what we mean by **AI** (Figure 3.3).

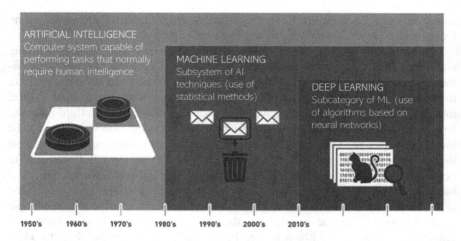

**FIGURE 3.1** Artificial intelligence and classification.

The term **AI** means the set of techniques, methodologies, and tools through which we try to "replicate" human intelligence. We can also say that AI means an "aggregate" of technologies, from ML to Natural Language Processing, which allow machines to perceive, understand, act, and learn. In other words, AI is a computer system capable of performing tasks that normally require the intervention of human intelligence. Most of these systems are based on ML (**Machine Learning ML**), while others are powered by deep learning (**Deep Learning DP**) (Figure 3.1).

Since AI "**allows**" to replicate human intelligence, it should be noted that these operations include planning, understanding language, recognizing objects and sounds, learning, and solving problems.

We often tend to confuse AI with ML and deep learning. In reality, **AI** means making a computer imitate human behavior in some way. ML, on the other hand, is a subset of AI or better yet a declension of AI implementation, which focuses on the ability of machines to receive a series of data and to learn on their own, modifying the algorithms as they receive more information about what they are processing. It consists of techniques that allow computers to understand things from data and provide AI applications. Deep **learning** is a subset of ML and indicates that branch of AI that refers to algorithms inspired by the structure and function of the brain called artificial neural networks. Deep learning, in other words, is part of a larger family of ML methods based on the assimilation of data representations as opposed to algorithms for performing specific tasks. With deep learning the learning processes of the biological brain are simulated through artificial systems (artificial neural networks) to teach machines not only to learn autonomously but to do it in a more "deep" way as the human brain knows how to do where with the deep term we mean "on multiple levels" (i.e. on the number of hidden layers in the neural network – called hydden layers). It is quite easy to understand that the more intermediate levels there are in a deep neural network, the more effective the result. Another very important observation is that while deep learning systems improve their performance as the data increases, once a certain level of performance is reached, ML applications are no longer scalable even if you want to add more training datasets.

**Google Brain** project, one of the best-known demonstrators of deep learning algorithms, has demonstrated how an artificial neural network can learn autonomously and hide the content of its messages during its behavioral interactions. In the experiment of Google Brain researchers, the Alice network had to send a secret message to the Bob network. Bob was to decode the message, and a third network, Eve, was to try and intercept it. Messages between Alice and Bob were found to be inaccessible to Eve. To achieve this result, Alice and Bob exchanged messages on the encryption key of the secret code, and after 15,000 attempts Bob was able to decode the encrypted message to go back to the original text. Eve, on the other hand, managed to guess only half of the bits that made it up. In other words, the experiment demonstrates that through ML, neural networks are able to use a simple technique to encrypt the information they exchange without having been programmed to do so, i.e. without having been supplied with specific data decoding algorithms. It is a result that has yet to be validated with other experiments, but surprising in principle because it supports the thesis according to which computers have the potential to learn by themselves. Today even a simple user, without in-depth IT knowledge, can try to interact with these systems using open-source software libraries available on the net. One of the specific frameworks for deep learning on which we ourselves have dabbled to implement experiments is **TensorFlow**. Open-sourced by Google, TensorFlow is an end-to-end open-source platform for ML. It has a comprehensive and flexible ecosystem of tools, libraries, and community resources that enable researchers to promote the state-of-the-art in ML and developers to easily build and deploy ML-based applications (as reported on the home page). However, the landscape of libraries used by the AI community tends to grow more and more. The fundamental aspect of all modern libraries for deep learning is certainly that of providing access to an automatic differentiation tool, so that the model definition part is greatly simplified (basically, the back-propagation is calculated by the library as efficiently as possible). Another essential aspect that characterizes almost all libraries is the interface with the GPU, in order to parallelize some aspects of the calculations. Historically, the first library to satisfy these two prerequisites and to gain great popularity was **Theano**, an open-source library developed within the University of Montréal. Theano library evokes the ancient philosopher, long associated with the development of the golden road, is a compiler optimized for manipulating and evaluating mathematical expressions, especially those with array values. This framework is historically applied in educational environments, but lately, it is giving way to new libraries. Microsoft Cognitive Toolkit (previously called CNTK), is an AI solution, initially released in 2016, which allows you to "train deep learning algorithms to operate like the human brain" (at least according to the framework developers).

Some of the features of the Microsoft Cognitive Toolkit include highly optimized components capable of handling data from Python, C++, or BrainScript, as well as ease of integration with other libraries such as Microsoft Azure and interoperability with NumPy.

Without going too far into technicalities, it should be emphasized that all these frameworks require knowledge of at least one Python or Java, or C++ language. Libraries on a third level have therefore been created, such as for example KERAS, which rely on the frameworks described above and simplify their use as well as facilitating the structuring of common models of neural networks.

### 3.1.3 Learning Models

Many ML problems can actually be classified as problems of minimizing a certain loss function against a certain set of examples (training set). The ultimate goal is to "teach" the model the ability to correctly predict the expected values on a set of instances not present in the training set (test set) by minimizing the loss function in this set of instances. The different tasks of ML are typically classified into three broad categories, characterized by the type of feedback on which the learning system is based: supervised **learning**, unsupervised learning, and **reinforcement** learning.

**Supervised learning** is probably the most frequently used ML in practice. These techniques foresee a training phase in which a dataset of "**labeled**" information is given as input to the ML algorithm. The algorithm will then operate by similarity on other unlabeled data. In this technique, both the input data to be analyzed and its results (output) are supplied to the computer. Wanting to exemplify the concept, let's imagine taking a sample of e-mails and adding a label on each one ("spam" or "no spam"). Figure 3.2 shows an example to estimate a general recognition rule.

Once the pattern is found, the machine uses it to classify all incoming e-mails into spam and no spam. In this way, I created a simple intelligent spam filter. This algorithm is called *the classification algorithm*.

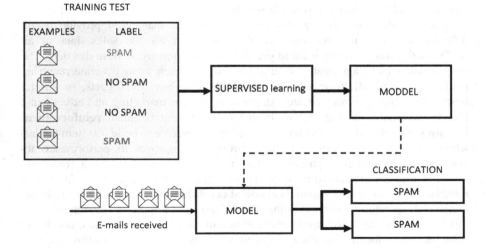

**FIGURE 3.2** Example of supervised learning.

One of the supervised learning experiments on which we have ventured more easily was that relating to computer vision. "Artificial vision" is a field of application of deep learning on which all those who have approached the topics of AI have tried their hand. Therefore, it has evolved considerably over the years. An artificial vision system consists of the integration of different components (from optical to electronic and mechanical ones) which allow to acquire, record, and process images. The result of the processing is the recognition of certain characteristics of the image for various purposes of control, classification, selection, etc. All the plutarchs of the new economy present by default the defaults in their applications, solutions based on computer vision. **Google**, for example, in the section dedicated to photos, catalogs the images, inserting them in pre-established categories managing to identify salient characteristics of the image itself. Our smart phone does the same. **Facebook** with the ability to recognize faces and tag them demonstrates how computer vision is our reality, in which we move more or less consciously. Finally, **Twitter** has the ability to recognize "illegal" images by deleting them instantly, without the need for a human supervisor, and so does **Instagram**. By virtue of their functionality and high calculation capacity, computer vision systems can find virtually unlimited fields of application, both in today's life and in the future (think of autonomous driving, and the consequent capacity they may have tomorrow these systems will reproduce "human sight"), both in the industrial field. Industrial applications undoubtedly include recognition of defects or tolerances, positioning and guiding robots, reading characters and product codes, contactless measurements, just to name a few. When there is an input dataset, but we don't have a defined output that we expect as a response then we talk about unsupervised learning. In unsupervised learning, as opposed to supervised learning, we have unlabeled data or unstructured data. With these techniques, it is possible to observe the data structure and extrapolate information. In these techniques, however, one cannot count on a known variable relating to the result or on a "reward" function. A classic example is when we have large data information about a population, and the algorithm returns recurring patterns that could provide useful information. Let's imagine that we have as input all the sales data in all **McDonald's stores** and we want to get some useful information from this data. The system manages to aggregate the data into clusters such as to generate recurring patterns, such as all those who buy BigMacs also buy McNuggets, or similar information aggregations that are extremely useful for marketing and advertising. Finally, when the machine learns from its mistakes, it is called **reinforcement learning**. The objective of this type of learning is therefore to build a system (agent) which, through interactions with the environment, improves its performance according to a try-and-error paradigm. This methodology provides for a constantly evolving training set according to a dynamic that allows you to learn from your mistakes. The example of **Pavlov's classical conditioning** lends itself well to better understanding reinforcement learning. One of the most important characteristics of this type of learning is the involvement of automatic or reflex responses, not due to voluntary behavior. It is a type of learning whereby an originally neutral stimulus, which does not elicit a response, is able to elicit one later thanks to the associative connection of this stimulus with the stimulus that normally elicits this response.

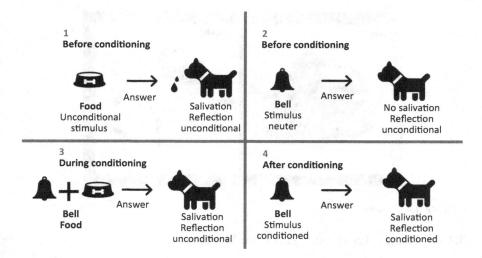

**FIGURE 3.3**   Pavlov's experiment.

This is exactly what **Ivan Petrovich Pavlov elaborated**[4] in his research. The phases of Pavlov's experiment are shown in Figure 3.3.

This system is the one underlying the functioning of this learning modality (unsupervised learning). In our experience, we have tested this deep learning algorithm on **COZMO,** a small robot produced by ANKI, which represented the progenitor of a series of domestic robots (Figure 3.6). Basically, Cozmo, with its unsupervised learning algorithms, is able to learn daily from its mistakes. While we worked on completing our text, Cozmo wandered around the desk, identifying obstacles, defining the perimeter as far as where to go so as not to fall off the table and looking for the best route to reach his favorite games. Thanks to the camera and its visual recognition systems, every now and then he crossed our faces, recognized us, and greeted us. An excellent tool for experimenting with AI and its possible applications. In fact, Cozmo is connected to the network, and every new teaching and every new game, he is able to transmit to similar devices connected all over the planet. There are several games that we have played with Cozmo that he probably acquired from other "colleagues" scattered in some other house since they had not been "suggested" by us (Figure 3.4).

Unthinkable goals have been reached over the years; new goals await us in the near future. One year ago, after the great success of ANKI Cozmo, its all-round improved successor, the ANKI **VECTOR**, was released on the market. In fact, Cozmo had very little computing power, and most of its functions were limited by the presence of an internet connection and APP connection. **VECTOR** is a greatly enhanced model and has improved in every aspect compared to its predecessor including microphones, touch, display, cameras, motors, anti-fallanti-fall, and collision sensors. It is also capable of interacting with ambient sounds, including our voice in a very realistic way. We have not yet met Vector as we are still too fond of Cozmo, but we have promised to include Vector in our team of "researchers".

**FIGURE 3.4**   Cozmo vs Anki.

### 3.1.4   GLOBAL TRENDS AND EUROPEAN STRATEGY ON AI

The different continents and states are following different strategies for the development of AI technologies; among these, China, the United States, and Canada are today the main protagonists at an international level. China has a government-led strategy, whose goal is to become the first developer of AI tools by 2030. In the United States, there is currently no government strategy: the government invests mainly in research for military defense, leaving it to large companies (such as Google, Facebook, Microsoft, and Amazon) to drive basic and applied research. In 2019, the States invested 6 billion dollars of public funding in research, of which 5 was for military defense. European states have activated individual national plans, such as the United Kingdom, France, Finland, Sweden, and Germany. Certainly, beyond the specific policies of each country, there is no doubt that the development of AI has become a central theme for all industrialized countries and emerging economies over the last few years. In terms of investment, the current effort by the United States and China to acquire dominance in the AI sector is far superior to that of other countries. In the United States, in February 2019, President Trump launched the new *American Artificial Intelligence Initiative*, which subsequently translated into a series of initiatives focused above all on sectors such as transport, agriculture, and meteorology. Alongside the two superpowers, industrialized countries such as Japan (as early as 2015), South Korea (2016), Canada (2017), and emerging economies such as India (in 2018) have adopted national AI plans. The proliferation of national plans has not escaped the European Community institutions, which have decided to strengthen coordination between Member States in the framework of the "**Digitizing European Industry**" program.

According to national policy, it is highlighted that States worldwide should concentrate their investments above all in five specific **areas:**

- *IoT, manufacturing, and robotics*. In automation and manufacturing the use of AI declined for the predictive and also prescriptive models, predictive maintenance, *digital twins,* and in general, for the improvement of production processes, potentially also with significant energy savings.

- *Services, healthcare, and finance.* In the financial sphere, AI developments could further consolidate this model, as well as make it more transparent and sustainable, through the development of more inclusive and customizable financial technologies (fintech). Even in the field of health, AI can support cutting-edge technologies on these issues.
- *Transport, agrifood, and energy.* As far as mobility is concerned, AI can support a more sustainable transport model. Agrifood is certainly a sector where AI, together with genetics and nanotechnologies, can continue to improve and guarantee both production and the maintenance of quality, also in this case by reducing waste and environmental impact.
- *Public administration.* PA can be the real driving force of growth, if modernized and equipped with adequate tools and skills to support an entrepreneurial fabric made up above all of micro-enterprises, and a citizenry that is still often unaware of the issue of digital technologies.
- *Culture, creativity, and digital humanities.* In cultural heritage, technology can help maintain and further enhance, both from the point of view of preservation and from that of improving the visitor experience.

## 3.1.5 BENEFITS AND RISKS OF AI

The development of AI is raising many questions about its benefits and risks. As for the former, they are now evident and important. In general, they go in the direction of transferring strenuous and repetitive human physical activities (an example is robots equipped with AI) and cognitive activities (tedious calculation sequences or surveillance/control of the correct functioning of machines) making them more efficient and safer; think of autonomous driving), but also in the possibility of creating new services (think of telemedicine and personal assistance) and new ideas that lead to new innovations.

The problems or risks in summary concern:

- the replacement of humans in many jobs (typically those that are not very manual) with consequent negative effects on employment (the latest Goldman Sachs report predicts that generative AI will lead to a loss of 300 million jobs);
- the difficulty of maintaining control of the decisions/actions/recommendations of an AI in terms of fairness and correctness, also in order to identify those, sometimes deliberately false, produced to influence people's choices/convictions and democratic life.

To face these risks, as has always happened with technological revolutions, it is necessary that governments, with the involvement of the various actors, implement actions so that the transition from the old socio-economic context to the new one is socially acceptable, respects the rights of people, and leads to new collective well-being.

For employment problems, it is important that governments develop suitable **welfare policies** and encourage the creation of new jobs by also investing in the training of workers.

For the **ethical ones**, special codes of conduct can be created (we also speak of algorithms), and the producers of applications can be qualified. At present, all these issues are under the attention of public and private institutions in the developed world, which have already taken various initiatives.

After the spread of generative AI systems, such as Chat **GPT-4 by Open AI** and Bard by Google, the discussion on the risks of AI has become particularly animated, and some authoritative players are hoping for a pause for reflection.

### 3.1.6 EXPLAINABLE AI (XAI) AND THE FUTURE OF AI

For some years now, in order to solve the problem of the risks associated with the use of AI, in addition to defining specific rules, an R&D effort has also been underway to develop techniques for realizing the so-called **human-centered AI**, i.e. the adoption of methods that the needs of the people and the society in which they live are at the center. One of these is the XAI (term coined by Van Lent in 2004) which aims to allow the user to know how an S-AI arrives at its decisions/actions/recommendations and how it will behave in the future. A theme that has become of great importance with the development of ML and in particular of deep learning which in fact constitutes a black-box approach based on complex models with a large number of connections and nodes.

In a certain sense, the XAI also aims to achieve the human-in the-loop, in order to balance the fact that an S-AI is more capable than a single human being in the specific application field (think, for example, how relevant a certain opaqueness of an S-IA can be in the medical, health, or military field). DARPA (U.S. Department of Defense – Defense Advanced Research Projects Agency) defined explainable AI as an AI capable of explaining its logic to a human user, characterizing its strengths and weaknesses, and conveying an understanding of how it will behave in the future.

Put this way, it is evident that explainability also includes interpretability, comprehensibility, correctness of the process, and transparency. It is evident that all this requires that the design team must also take into account the multidisciplinary nature of AI and in particular the human psychology of explanation, which draws on the vast research field of the social sciences: to explain something one must take into account the level of understanding of the interlocutor and his/her aims. The HCI (Human Computer Interface) interface itself must be evolved and consider the voice, and the facial/gestural expression.

At present, many XAI techniques produced by various universities and public and private research centers are being tested. The DARPA studies concluded at the end of 2021 analyzed many techniques by focusing on the various problems still to be solved.

Among the difficulties that have emerged are those of providing explanations customized to the type of user, maintaining high performance, and above all, evaluating the quality of the "explainability" performance which, as previously mentioned, includes various aspects. The path taken, therefore, still requires R&D

work and, while not solving all the problems at hand, represents an important step forward in the future of AI.

Definitely, autonomous AI systems are now an ever-developing reality that are revolutionizing our society. A revolution that, as has happened other times in human history, cannot and must not be opposed but must be governed in order to make the most of its benefits and minimize its negative effects. Governments are right to put in place appropriate regulatory actions and policies to accompany the resulting socio-economic transformation, even if it is a difficult task.

In this context, the development of XAI appears to be an effective technological solution to solve the main problem of the opaqueness of the new S-AIs in terms of understanding, transparency, correctness of decisions/actions/recommendations, also pursuing the objective of human-centered AI with man entering the Loop. This is a necessary challenge for sector R&D to which all players in the field must contribute.

## 3.2   VIRTUAL ASSISTANTS, SOCIAL NETWORKS, AND CHATBOTS

According to the **2023 Digital Report**, as regards new technologies, we register an interesting level of maturity. In fact, the use of voice assistants from mobile phones or dedicated devices grew by 5% points compared to last year (from 30% to 35%).

It is certainly interesting to note that the preferred platforms are **YouTube and the Facebook** family of apps (WhatsApp, Facebook, Instagram, and Messenger, in order). And **Instagram is** the platform that has recorded the most evident growth in recent years, from 55% to 64%. Other platforms that are starting to emerge in use by citizens are **Snapchat**, **Twitter**, **WeChat**, and **Reddit**. They are all very different platforms, and it is interesting to note how the growth is widespread, an indicator of the "general health" of the category and not just of some specific niches. The use of **TikTok** also emerges which naturally is an indicator of an even wider diffusion on the demographics predominantly present on the platform, that of **Gen Z**, or the generation that follows the Millennials, generally confined between the second half of 1995 until the year 2010. An important aspect of this generation is its widespread use of the Internet almost from the earliest years of life. A very important fact emerges from the report: the use of **voice commands** is widespread in all countries, but China appears at the top of the ranking, followed by Indonesia and India. Bottom-ranking countries are Japan (17%) and Denmark (16%).

Another interesting fact is that **conversational systems** are spreading more and more. Through them, it is possible to ask questions although they don't have enough information to answer. Arguably, the future of work will be a hybrid system involving both human and machine intelligence. In this scenario, the two entities will work together to achieve shared goals and will interact with voice and natural language. There are several examples of conversational systems in social fields, such as the **conversational system to support triage, a chatbot** that supports healthcare personnel in identifying the clinical condition of patients. As the use of intelligent conversational systems becomes common, biometric voice recognition is also gaining strategic value in a wide range of sectors including financial services, retail, public administration, health, and safety, as well as telco operators and service providers. The identification of people through voice promises to increase the level of personalization of the services provided by contact centers both traditional with operator and automated with voice recognition. However, in order to use voice biometrics it is important that specific solutions, on premise or cloud, can integrate, on the one hand, with business applications and any conversational systems, and on the other, with other security systems that can strengthen the certainty of the authentication. Typical areas for using voice biometrics are **contact centers** to authenticate callers, reducing customer recognition time. Other applications concern, for example, **workforce management**, replacing traditional badges, or the possibility of **creating voice signatures** with which to sign insurance documents or authorize financial transactions. A peculiarity of biometric data is obviously the protection of **privacy**. In any case, biometric technology is establishing itself more and more and Amazon itself launched **Amazon One** (https://youtu.be/xH_SVNVIfzk) in 2020, the biometric recognition technology that will allow customers to pay simply by touching a sensor with the palm of the hand.

Using Amazon Go it is not necessary to go to the till because the shopping bill will be calculated independently by a smart network of sensors and video cameras that record everything you put in cart. The first time the customers scans their hand and associates them with the credit card, the next time the system can already be used. The hand therefore becomes a sort of **credit card**, taking advantage of the fact that humans have different and unique hands (i.e. shape, superficial details such as the lines up to the veins under the skin, etc.). In any case, users will have the option of deleting biometric data from a portal if they no longer wish to use the service. Amazon plans to extend the technology to third parties as well, for example, for tickets in stadiums and of course, allow payments in other shops as well. It is not a simple payment technology, but it is an **identity technology**. The decision to opt for palm recognition instead of other options such as facial recognition is due to some privacy advantages. One reason is that palm recognition is considered more private than some biometric alternatives because it is not possible to determine a person's identity by looking at an image of their palm. The model obviously therefore represents one of the next frontiers of the store of the future.

### 3.2.1 Talking to Computers: The Era of Virtual Assistants

In the last ten years, and mainly thanks to the recent advances in AI, the technology of natural language conversation with computers has made great strides. Speech recognition, understanding and generation of natural language, speech synthesis, and dialog management have reached levels of performance such as to make it possible to market virtual assistants, including Siri, Alexa, and Google **Assistant.** Virtual assistants have been a technology used in industrial applications (e.g. customer service) since the mid-1990s. However, only today that we see their widespread use in the consumer market, with a great opportunity to improve and simplify, at all levels, interaction with the digital world and the vast amount of information on the web, domotic control, and communications. We believe that in the near future virtual assistants will be increasingly integrated into our private and professional lives. In our opinion, it is a crucial issue for digital and technological development.

In this regard, it is necessary to analyze what is meant by natural language and what are the next challenges in the world of "voice". It is of considerable importance to specify immediately that the **voice is a particular technology**! The world began to realize that computers could talk in the 1950s with the first experiments done at Bell Laboratories. Today men can communicate with computers and with the vast world of the Web, apps, and various services since today computers are able to understand the voice and the meaning of what is said and therefore process an answer. Even if the world of voice communication with computers has made great strides in the last 30 years, deep learning techniques, which have appeared on the scientific/technological scene for a decade, have demonstrated performances never seen before for the resolution of various problems, and among these voice communication with computer. Technology is very close to the **possibility of replacing the keyboard with the voice**. Traditionally, VOICE has been used primarily in the **service sector**. The question is how voice systems could influence other sectors, such as industrial production? Well, the answer is not simple. It really depends on the corporate vision. Surely, in service companies, the usefulness of voice recognition is intuitive since there is a reduction in costs associated with human resources. However, people don't always like to interface with a computer. Voice recognition is currently very reliable, just think that out of 100 questions, generally 80% of the answers are correct. Samsung recently unveiled a *very sophisticated frontier avatar of the future virtual assistants*: **NEON**. The avatar is described as an *artificial human being* able to *converse and sympathize* like a real person. This is not a robot or a virtual assistant like Siri or Alexa, but a personalized digital figure that can appear on a display or a video game, designed to be a playmate, an actor, or even a TV presenter "learning, evolves and builds memories from interactions". According to the tech giant, Neon will be able to give answers in fractions of a second and therefore also be available for services to companies to improve customer service interactions and make customers feel like they have a "friend". This is just one of many examples of technologies that attempt to embody assistants, both virtual and physical, such as the first social robot **Jibo**.[5] While, Jibo had somewhat limited goals, examples such as NEON to create an avatar of a human being are somewhat perplexing. side of the media "wow" effect, these systems, still under development, foreshadow very interesting potential. It is therefore natural to ask how mature the voice/AI combination is. We know well, as Judea Pearl explains in her text *The Book of Why: The New Science of Cause and Effect* that human intelligence can have three levels of representation: first associative level, second level of inference or predictive ability, and third level of "counterfactual" or making inferences about the past, for example *what would have happened if ...* . That is, human intelligence of the highest level that allows us to develop theories. Today's AI is only at the associative level, that is, it allows us to make associations, such as giving voice to words of text, or from words of text or their meaning. We are still far from the second level, the inferential one, and light years away from the "counterfactual" one.

At this point, we might ask ourselves what we could expect from the **near future**. Surely, we will see an evolution of **communication between man and machine**. The computer will have to be able to communicate "naturally" with a human being, and to do this it will be necessary to further develop natural language processing techniques. But beyond that, computers will be able to support us in our daily tasks which take up a good deal of our time. Probably, in the near future, we must expect that there will be some "solution" that will replace ourselves, in support operations, a sort of **"digital butler"** that will help us, for example, to organize a dinner with friends, shopping, or planning a trip. Indeed, we are certainly more interested in going to dinner with our friends than wasting time organizing the dinner itself.

## 3.3  QUANTUM COMPUTING

Quantum computers are a "wonder box" and feature subatomic particles such as photons or electrons, which can store much more information than current computing devices can. The first time we encountered a **quantum computer** was in 2018, during the **Consumer Electronics Show**[6] 2018 in Las Vegas (Figure 3.5).

On that occasion, during the keynote speech by Intel CEO Brian Krzanich, which opened the event in 2018, the first 49-qubit quantum processor ready to be marketed was presented.

Honestly, the object presented and displayed like a star looked anything but a computer as we understand it today. After all, who could have imagined that the set of music devices, cameras, telephones, PCs, and much more could be enclosed in a

**FIGURE 3.5**  IBM quantum computing.

smart phone just 60–70 years ago? After this experience, we started to read up on it to better understand what it was about aqnd what its characteristics and potential were. The first reflection that led us to delve into the matter in question was a sentence by **Richard Feynman**[7] from 1982, in which he very simply stated that nature is governed by the laws of quantum mechanics, and therefore in order to be able to explain natural phenomenologies, there is there was a need for a computer that "reasoned" with the same laws. A lot of progress has been made since then, and in a few years the first processor at the base of quantum computing was born: the **QUBIT**.

**What differentiates the qubit from the bit?**

Bits are binary systems, therefore, as we know they can only assume two on/off stages, one state rather than another. For example, if we wanted to know our position on earth using a single bit, the only information we could receive was whether we were in the southern hemisphere or northern hemisphere. Physically in computers these 0 and 1 exist due to the effect of transistors, devices that we can imagine as microscopic switches which, when closed, return 0 and 1 when open. The electrical energy passing through these circuits is affected by these transistors. The qubit, on the other hand, can provide us with much more detailed information. Returning to the previous example, a qubit could give us back a precise position on the Earth's sphere. This is because the "logic" with which a qubit acts is not binary, but *"probabilistic"*, i.e. both the states 0 and 1 can be present at the same time. The laws of quantum mechanics, in fact, postulate that every particle is subject to the so-called **superposition principle**, i.e. each particle can be in several different states at the same time, with different probabilities. The qubit, with respect to the bit, we could represent the state of a subatomic particle (an electron or a photon). As we know this particle, for example, the electron, in addition to rotating around the nucleus, also rotates on itself (the spin). Well, we could say that if this electron is in an electromagnetic field, this field influences its direction, and therefore we could say that the value will be equal to 1 (spin up) if it moves in one direction and 0 if it moves (spin down). In the other, and in quantum physics the two states can be present simultaneously until selected. This is what a qubit can do, i.e. the possibility of representing both the state 0 and 1 at the same time. It is precisely thanks to the superposition principle that it is possible to overcome the on/off dualism and to convey much more information. Ultimately, the qubit allows you to parallelize calculations, that is, to carry out many, many operations simultaneously. In this regard, we recall the mental **experiment of the cat by E. Schrodinger**,[8] according to which a cat present in a box with a poisonous gas, until the box is opened, can be both alive and dead (it has a 50% probability for one state or the other). Therefore, until we go to verify the state of the cat we must assume that it is simultaneously alive and dead (quantum superposition). When we go to verify the condition of the cat in the box, at that point, our observation will have forced the decision and therefore the animal will be either alive or dead. From this example, it follows that if we put two bits in a series we can have up to four pieces of information (00–01-11-10). And of these four we can only receive one. A qubit, on the other hand, thanks to the quantum superposition property, can represent two pieces of information at the same time. If we use two qubits we can have four pieces of information at the same time. And so on, each addition of a qubit generates an information increase equal to 2 raised to the power n, with n equal to the number of qubits we are using. In addition, to the superimposition of states, qubits have other specific properties that derive from the laws of quantum physics such as ENTA-NGLEMENT,[9] i.e. the correlation between one qubit and another, from which a strong acceleration in the calculation process derives.

This property suggests that there is a strong correlation between the different quantum quantities, between the different qubits, allowing us to measure the information contained in one of these quantities or, if we want, in a qubit and determine from this measurement the information contained in the adjacent qubits. In this way, it is possible to monitor an impressive amount of data. Thus, in essence, this quantum phenomenon allows you to determine the quantum state of a specific physical system, through the superimposition of multiple systems. Observing one system simultaneously determines value for the others as well, regardless of where they are located. Thus, it can be demonstrated that entanglement implies the presence of correlations at a distance (theoretically without limits) between their physical properties.

A team of physicists from the University of Glasgow in the United Kingdom has shown the first image of entanglement, the *strange interaction* between particles that underlies the phenomenon and functioning of quantum computers.

To better understand the potential of this calculation tool, we report an example. Let's imagine we want to organize a trip and want to visit, moving with an airplane, tencities one after the other, as quickly as possible and with the lowest possible cost. A traditional computer will provide us with an optimal itinerary in 20/30 seconds. If the number of cities goes to 15 or 20 the computer will need about 10,000 hours of calculation. If we increase the number of cities to 35, the calculation will involve a time equal to the age of the universe. The use, instead of the qubit, would allow us to solve this problem in a few seconds. It is therefore understandable how the potential of quantum computing is enormous. The importance of the topic is demonstrated by the awareness that although the era of quantum computing was inaugurated in 2001 by IBM, there are currently many research centers and companies around the world investing in the development of quantum computing. In Europe, the European Commission has launched and financed the *Quantum Flagship Initiative* project, with the primary objective of accelerating the development of quantum technologies in the European Union. In the United States, the **National Quantum Initiative program** has been proposed in which the National Science Foundation, the Department of Energy and Fermilab of Chicago participate. Italy is the only country called to collaborate on the project by being awarded a loan of 1.5 million dollars. A similar plan is also being conducted in China. Companies like **Amazon** are increasingly investing in the quantum business by offering select enterprise customers the ability to experience early-stage quantum computing services in the cloud. It's called Amazon Braket, and it's a platform-as-a-service designed to help companies take advantage of quantum computing by developing and testing quantum algorithms in simulations. Amazon isn't investing to build its own quantum computer, it's putting its computing and software resources into programming it, in the future, when everything is ready.

According to IBM, the new great frontier of information technology will go even beyond the simple use of quantum computing, but will combine *bits* + *neurons* + *qubits*, i.e. *supercomputers* + *AI* + *quantum computers*. This integration will make it possible to take advantage of what could be defined as "accelerated research", which is already applied to various sectors today. In fact, with the entry of quantum computing the limits of Moore's law will be overcome, and there will be an exponential growth of computational capacity. In the future, technologies such as cloud computing, high-performance computing, and AI will be closely linked to quantum computing and will generate value for businesses, the scientific community, and society as a whole.

### 3.3.1 QUANTUM POTENTIAL

Let's try to understand what all this computing power can do for us. Let's imagine, for example, that we want to simulate a water molecule ($H_2O$) with the binary system. Well, to "reproduce" the molecule with chemical precision, 10,000 bits will be needed, which seems a lot, but much less than those needed to transfer a photo to our smart phone. If we take a more complex molecule, such as that of $C_2H_6O$ ethanol, the number of bits needed to be able to simulate it will be $10^{12}$ (1,000 billion bits). For even more complex molecules like caffeine, it will take $10^{48}$ bits. In other words, it would take a computer half the size of the moon to make this calculation. If we analyze molecules such as penicillin ($10^{86}$ bit) or sucrose ($10^{82}$ bit) for these the calculation would require the creation of computers that would reach such dimensions that they cannot be contained on Earth. In fact, these numbers are far greater than the atoms in the universe.

*But what does it mean to* "simulate" *a molecule?* Think of **Star Trek**, teleportation, and dematerialization. While human teleportation exists only in science fiction movies, in the world of quantum mechanics, teleportation is possible. In the quantum world, in the current state of knowledge, teleportation involves the transport of information and not the transport of matter. However, quantum computers are advancing rapidly, and there could also be important improvements in the field of matter transport. Furthermore, through quantum computing, we could think about the possibility of identifying the right mix of molecules capable of combining with our DNA and eradicating pandemic diseases. High-performance computing through supercomputers and AI can help accelerate this process by rapidly screening billions of chemical compounds to find relevant drug candidates.

In any case, perhaps to better understand the phenomenon and its scope, it is useful to bring an example closer to our daily lives. If we think of the operation between two prime numbers, simple as 5 and 7, even without the support of a PC, we know its result well, and in the same way, we also know how to do its inverse operation, i.e. the decomposition into prime numbers. Well, however, if we increase the number of digits to be multiplied, however complex, the multiplication operation can be performed by any computer. The reverse action, however, for a workstation even of significant dimensions is not at all simple, indeed very, very complex.

18462047204702479127437097170932809471029790874983709183609811986

×

878298734567890345679876534567890345678098765431234567789234567899

The following multiplication, for example, is "simple" for a calculator, while the reverse calculation would take several hundred years. Binary computers can do it, but in an inordinate amount of time. By virtue of these considerations, cryptographic systems are born. These systems are those which, through "keys" held partly by a user, and partly by our bank branch, allow us to preserve our credit cards, current accounts, and everything that revolves around finance. In fact, the decoding operation, aimed at decrypting our secret code suitable for withdrawing from our ATM, is nothing more than an operation of the type described in the previous example. Therefore, a normal (or even sophisticated computer) to be able to decrypt our "access keys" would take a very high number of years, and in any case far longer than the expiry period of our credentials and of the card itself. This is the system that makes our current accounts "safe". A quantum computer instead, with a specific program, could do it in seconds. The decoding program already exists; indeed it has existed for several years, created by **Peter Shor** of AT&T in 1994. Shor's algorithm was created to be able to decompose any number into prime numbers. In his day Shor to factorize a 129-digit number, had 1,600 workstations work in parallel for 8 months. To factorize a number of 250 digits a binary computer would take 800,000 years, for a number of 1,000 digits it would take a number of years greater than the age of the whole universe. Shor demonstrated that with a quantum algorithm and a machine capable of using it, the operation would have required a few million steps.

### 3.3.2 QUANTUM SUPREMACY

A study recently published in *"Nature"* (following a document that appeared in the Financial Times and then withdrawn) reported the news that the research team led by the **Google Quantum AI Lab** had managed to achieve what, **John Preskill**[10] in 2012, defined **quantum supremacy**. More in detail, the article reported that the digital giant managed to perform an extremely complex calculation in 3 minutes and 20 seconds (just 200 seconds) using its 53 qubit Sycamore quantum microprocessor (actually was made up of 54 qubits, but one didn't work). Google scientists claimed that IBM's SUMMIT supercomputer, the most powerful in the world today (the size of a basketball court) would take at least 10,000 years. An unequivocal affirmation of the achievement of quantum supremacy (quantum supremacy), i.e. being able to beat the performance of the most powerful traditional computers around. In truth, IBM did not agree with this result at all and replied, after a while, with a post on its site that a different configuration of the supercomputer, with additional storage capacity, would have made it possible to resolve the operation (finding recurring patterns in a random series of numbers) in a maximum of two and a half days. And more precisely. In our opinion, however, two and a half days equals 216,000 seconds, a few orders of magnitude more than Sycamore. Briefly, this test that determined the supremacy of a quantum computer over a binary one consisted of solving a random number demonstration problem (RCS – Random Circuit Sampling), creating combinations of random values that took into consideration all 53 qubits (that is, 2–53, a huge number of calculations). Quantum supremacy essentially consists in solving a quantum problem, which is difficult to solve by a classic electronic processor, in acceptable times. This "discovery" certainly does not provide us with elements regarding the times of use of quantum computers in our lives and at the same time does not imply their generalized use like that of our current PCs. In other words, one must not think that the quantum computer will soon replace traditional calculators for conventional operations. That would make no sense given the large amount of energy required for a quantum computer to operate. In all likelihood, the use of such a complex machine will be oriented and specific to sectors, such as, for example, materials science, the pharmaceutical industry, tumor medicine, or particle physics. In these scenarios, a quantum processor could really make possible technological advances of very far-reaching and difficult to predict a priori. In fact, quantum computing, although in its infancy, already exists and is used: from financial models to medicine, including research for COVID-19, from weather forecasting to cryptography. In this regard, it should be noted that quantum computers have the potential to overcome all the barriers of traditional cryptography. Quantum **cybersecurity** will be the basis of quantum communication networks, combining power and speed with security. However, there are also some **problems** related to the use of quantum computing. The first challenge is to maintain the quality of the qubits, which lose their special quantum qualities rapidly (in about 100 microseconds) due to factors such as vibrations and fluctuations in ambient temperature and electromagnetic waves. Precisely because of the temperature factor, to guarantee the functioning of a quantum computer it is necessary to reach very low temperatures close to absolute zero (about −273°C). To obtain them so far the most common method has been the use of liquefied gases (such as the helium-3 isotope), but it is a very expensive system. The next frontier of research will also be oriented towards how to allow the functionality of these devices, reducing the amount of energy to be used.

## 3.4 BLOCKCHAINS

If we wanted to find in history a first example of logic comparable to that which supports the **blockchain**, we can probably turn our memory to the island of **Yap in Micronesia**. The local populations used a stone as a currency. A particular stone. We are in 1400 when the inhabitants of this small Micronesian island discover particular rocks on the nearby island of Palau. These rocks, which we know today under the name of **RAI**, appear as circular discs dug into the limestone, with a large hole in the center. The size of these stones varied from a few centimeters in diameter to those that weigh almost four tons and have a diameter of three meters. The value of these stones, however, was not linked to their size, but to the history, sometimes handed down, by the stone itself. If, for example, many people had lost their lives in transporting it, or the method of extraction of the same had been particular, or even if the stone had been lost in transport from Palau to Yap, the value that stone assumed was left unchanged, even if the stone was no longer physically present. The same stones, in passing from one owner to another, by marriage, inheritance, or simple exchange of food, were often not even moved to the homes of the new owners. The method of transferring the property took place with a collective acknowledgment of the transaction, handed down orally, and recorded in a public register that all the inhabitants owned, which was updated after each single transaction. Well, blockchain technology is essentially based on the same principle. In fact, the blockchain (literally a chain of blocks) is a distributed database resistant to possible tampering which continuously maintains a growing list of distributed ledgers, i.e. systems based on a distributed ledger (just as the value and the possession of RAI stones), which refer to previous information. The blockchain is made up of blocks of transactions correlated by a timestamp, the so-called timestamp, i.e. a number that expresses a time quantity; more precisely, the timestamp corresponds to the number of seconds that have passed since a conventional date known as the Unix Epoch (January 1, 1970). Like the Yap Island Registry, all updates are publicly known, in this case, handed down with data, not verbally. But the analogy makes us understand how all this happens without the guarantee of a third party, a bank, a state, or other intermediary. Each block includes the hash (blockchain output values, known as **hashes**, are used as unique identifiers for data blocks) of the previous block, which is a type of control system that links blocks together. The resulting structure forms a chain in which each additional element verifies and reinforces the previous ones. The transactions are transmitted and entered into a collective control system that considers the longest chains or chains that are computationally more difficult to forge valid. The consensus reached between the nodes thus eliminates the need for third parties or central hubs to verify the transactions and exchanges recorded via the network. A declination, of the use of the blockchain, and a trading currency with the same characteristics as the RAI (even if not material), or a **cryptocurrency**. The best-known cryptocurrency is **bitcoin**.[11] The bitcoin source code was first released by **Satoshi Nakamoto**[12] (to date we don't know who he is or if it's just a fantasy name; theories about his true identity are numerous. No one knows if it's a "he", a "she" or if it is multiple people) on **January 9, 2009**, in version 0.01, on **surface forge**, a developer community and shared space for open source software. The construction of the bitcoin architecture does not arise in a vacuum, but arises from a stream of reflection articulated on the intersection of information technology, cryptography, and finance. Some parts of the bitcoin system and the blockchain are "daughters" of reflections made by other researchers in the past, such as the concepts of **Proof-of-Work (PoW)** (consensus algorithm at the base of BC) and **hash cash** (represents both the mathematical problem that its solution in the blockchain/bitcoin system) elaborated by **Adam Back**, or the very concept of **cryptocurrency,** based on a decentralized **ledger**, or ledger, first proposed by Wei Dai in 1998. Furthermore, in past the ground had already been seeded with theorization of cryptographic coins such as **David Chaum's**[13] **digicash** and **Nicholas Szabo's**[14] **bit gold**. It should be noted that the latter is considered by many to be the most probable of the individuals known to the community to be the natural person behind the phantom Satoshi Nakamoto: his pervasive knowledge of the sector, his almost total disappearance just before the publication of the original paper, even the similarities in prose and terminology make one of the more likely candidates. The name Satoshi continues to hover in the community and on the markets despite its silence, but perhaps probably precisely because of its silence. Respect for the founder is so high that the bitcoin community has decided in his honor to baptize the minimum unit of account into which a single bitcoin can be divided, i.e. one 100 millionth of a bitcoin, "satoshi". Wanting to simplify, we could say that bitcoin is built on three bricks:

- the public transaction register, called **cryptoledger**, i.e. a list of exchange actions carried out by the various addresses within a system protected by robust cryptography;
- the structure of transactions based on the so-called robust double-key cryptography; users have a private key that allows them to spend money and sign transactions, and a public key that allows them to verify the signature and impersonate the bitcoin wallet address;
- the blockchain; transactions are packaged by the network in linearly connected blocks, with a collective and repeated verification system that allows longer chains to be considered valid, which, at the algorithmic level, are more difficult to forge.

Bitcoins are periodically created from scratch with a method called **mining**, a process linked to the intrinsic functioning of the block generation and transaction validation system. A fixed number of bitcoins is released approximately every 10 minutes, with an assignment rule that assumes the resolution of very complex calculations. Efficient mining is out of the reach of ordinary users today: the process requires specialized hardware and high investments. Those who want to obtain bitcoins can do so in more traditional ways, for example, by exchanging euros or dollars or by selling products and services. To deepen the analysis it is necessary to summarize the history of the idea related to **cryptofinance** and the **cultural milieu** in which these concepts have developed. Since ancient times, money has always been associated with something physical, which can be weighed and touched. However, the modern financial system has allowed us to move away from physical cash. Indeed, today it has become normal to use sales platforms such as **eBay** (launched in 1995) where all payment transactions are processed by the **PayPal platform** (established in 1999). Without considering that **digital banking** has become universal as has **mobile banking**, thanks to the diffusion of cell phones. Companies like **Square,** Inc. that offer financial and payment services through mobile devices are establishing themselves more and more. Square Jack founder **Patrick Dorsey**, best known for being the creator of Twitter, aspires to revolutionize the way money is used and that in the not-too-distant future all payments can be made via mobile. With the spread of digital payments, trust has become a crucial issue. Furthermore, another concept that has been establishing itself in recent decades is also linked to this phenomenon, namely, that of the sharing **economy,** declined in various ways and applied to various economic and social sectors. The first example of a sharing economy in the Internet age is indicated by many precisely in eBay. The phenomenon is recent, and the conceptual area to which it refers is vast and varied. Thus a series of contiguous, analogous, or parallel definitions have developed: from peer-to-peer economy to sharing economy, from gig economy to on-demand economy up to sharing consumption. In recent times, the phenomenon of **sharing mobility** has been gaining ground, which we could consider a subset of the sharing economy: the possibility of moving from one place to another through shared means and vehicles such as car sharing, bike sharing, scooter sharing, but also car pooling and similar ways of sharing. With the introduction of these economic changes, new rules have been established, codified algorithms that are gradually taking the place of "traditional" rules. Bitcoin was developed precisely in this new market. A strange object, which scares governments and large companies, but is growing exponentially. **Mark Andreessen**, the founder of Netscape in the 1990s and now one of the most famous venture capitalists through the Andreessen-Horowitz fund, argues that the blockchain is like "the **Internet in 1994**". Obviously, there is no shortage of criticisms and perplexities. Between 2013 and 2014 bitcoin got into the media with increasing frequency, often not in a good way. There has been talk of speculation, scams, the collapse of some hubs such as **Mt. Gox** (it was a cryptocurrency exchange site) and online black markets such as Silk Road. Even influential thinkers in the field of finance, such as economist Nouriel Rubini and former Federal Reserve chief Alan Greenspan, have expressed doubts about the system, associating it with a Ponzi scheme or a bubble. At the same time, bitcoin is attracting interest from investment banks, central banks, public regulators, and venture capitalists. Incubators (such as BoostVC and DeCentral) and investment structures (such as Pantera Capital) specializing in cryptocurrencies and the blockchain technology that underpins the bitcoin network are emerging. The bitcoin network has reached a planetary diffusion. Bitcoin was the first system capable of growing on a planetary scale, thanks to very ingenious technological solutions constantly perfected by the network of developers. The **rigidity** of the bitcoin network has led to the construction of dozens of other cryptocurrencies. Only 21 million bitcoins will be able to be created over time, through a linear curve predictable at the algorithmic level and a rigid monetary policy. By January 2015, more than half of the available bitcoins had already been created: about 13.8 million. The 21 million ceiling is not necessarily a limit in terms of its use as a unit of account: after all, 2.1 quadrillion (2.1 trillion) satoshis will be available. The problem is the lack of flexibility in monetary policy: the lack of discretion. Indeed, it helps to amplify the great volatility of bitcoin with respect to the main currencies (dollar, euro) or better still with respect to a basket of standard goods that can act as an anchor to measure their value. You may be wondering how much is a bitcoin worth. Due to the high volatility, the answer to this question changes dramatically over a short period of time. As of April 8, 2019, one bitcoin was worth 4,667 euros, after a long period of decline which saw the bitcoin-Euro price lose more than 80% of its value in the last year from the highs of the end of 2017. After a stable start, in 2019 saw a recovery of the world's most popular cryptocurrency – since the beginning of this year, bitcoin has already depreciated by almost 40%, a trend that looks promising for the value of BTC against the euro, collapsed over the past year. The logic behind

**FIGURE 3.6**   Bitcoin conversion to Euros and US dollars.

the value of a bitcoin in euros is the same as that found, for example, in asking how much is a pound in euros is worth: currencies have value in relation to the other currencies with which they are exchanged. Therefore, as with any other currency exchange, the value of bitcoin in euros changes many times even within a minute. Therefore, the answer to how much a bitcoin is worth in euros is not unique but must be researched and contextualized every day. Figure 3.6 shows an example of converting bitcoins to Euros and US dollars over time.

We are in a historic moment in which this innovation is attracting a great deal of attention. Many opinions about bitcoin are changing with time and understanding of blockchain technology. Probably the most interesting aspect related to bitcoin is not so much that of being a currency but a protocol, thanks to the potential offered by the block chain. Blockchain technology is part of a complex and constantly evolving universe that can be defined as the "**Internet of Value**", i.e. those systems that make it possible to exchange value on the Internet with the same simplicity with which information is exchanged today. Just as html has become the standard markup language for the web, the blockchain could have the same potential – become the protocol for reliable transitions. Consider, for example, transactions that require an intermediary as guarantor for the exchange of legal documents, brokerage commissions, and the purchase of tickets. Over the years, several cryptocurrencies have arisen that have tried to compete with bitcoin, correcting some "defects" of this electronic currency. Certainly the most widespread is Ethereum (ETH), released in 2015, which involves the execution of a form of "highly programmable digital money – smart contracts".

Also worth noting:

- **Ripple** (XRP), a decentralized cryptocurrency native to the RippleNet payment platform, was created in 2012 to offer banks and financial institutions a real-time gross settlement system, enabling secure and instantaneous financial transactions that can span the globe.
- **Litecoin** (LTC), which compared to bitcoin processes a block every 2.5 minutes (against 10 minutes for bitcoin) and produces scrypts in the execution of the proof-of-work.
- **Waves** (WAVES), the cryptocurrency that is the fulcrum of a project that envisages a decentralized monetary exchange (both FIAT money and cryptocurrencies), i.e. based on the p2p network. Currently, through the Waves-NG protocol, it is the fastest decentralized blockchain, being able to handle 1,000 transactions per second against 7 per second for bitcoin.
- **IOTA** (IOTA), a next-generation cryptographic token, was created to be lightweight and be used in the automatic payments that smart devices will make in the Internet of Things. IOTA is programmed in ternary and has a tangle instead of a blockchain, which makes it infinitely scalable and removes the need for transaction fees.

In addition to these just described, there are many others, but, honestly, we do not know if and how long these coins currently present on the net will last. In our opinion, those that will create greater "**encrypted**" trust as a guarantee for all users will certainly be those that will have the greatest probability of "survival". Today, several companies from all over the world have started experimenting with blockchain and distributed ledger solutions. The most advanced sector is certainly the **finance and insurance sector**, which was the first to take action to respond to the threat brought to it by bitcoin and which is already moving toward an application development phase of projects. There is no shortage of projects and applications in the **agrifood**, **advertising**, **logistics,** and **public administration fields**.

## NOTES

1 Kai-Fu Lee (born December 3, 1961) is a Taiwanese-born American computer scientist, businessman, and writer. He currently lives in Beijing, China.
2 Alan Mathison Turing (London, 23 June 1912–Manchester, June 7, 1954) was a British mathematician, logician, cryptographer, and philosopher, considered one of the fathers of computer science and one of the greatest mathematicians of the 20th century.
3 John McCarthy (September 4, 1927–October 24, 2011) was an American computer scientist who won the 1971 Turing Award for his contributions to AI.
4 Ivan Petrovich Pavlov (Ryazan, September 26, 1849–Leningrad, February 27, 1936) was a Russian physician, physiologist, and ethologist, whose name is linked to the discovery of the conditioned reflex in dogs, which he announced in 1903.
5 Jibo is the first social robot capable of watching, listening, and developing new skills. Jibo is currently no longer produced. It was powered by facial and voice recognition technology, so it remembered people and built real relationships with everyone it met (https://youtu.be/MNzb4FC6lhg).

6 The International Consumer Electronics Show is a consumer electronics fair staged by the Consumer Technology Association in the United States of America since 1967. It is held once a year, in January, at the Las Vegas Convention Center (Las Vegas, Nevada).

7 Richard Phillips Feynman (New York, May 11, 1918–Los Angeles, February 15, 1988) was an American physicist and science popularizer, Nobel Prize in Physics in 1965 for the elaboration of quantum electrodynamics.

8 Erwin Rudolf Josef Alexander Schrödinger (Vienna, August 12, 1887–Vienna, January 4, 1961) was an Austrian physicist, of great importance for his fundamental contributions to quantum mechanics and in particular for the equation named after him, for which he won the Nobel Prize in Physics in 1933.

9 Quantum entanglement is a quantum phenomenon whereby under certain conditions two or more physical systems represent subsystems of a larger system, whose quantum state cannot be described individually but only as a superposition of several states. The phenomenon occurs when two particles are intrinsically connected, and this union has effects on the physical system: any action or measure on the first has an instantaneous effect on the second too (and vice versa) even if it is at a distance.

10 John Phillip Preskill (born January 19, 1953) is an American theoretical physicist and professor of theoretical physics at the California Institute of Technology, where he is also director of the Institute for Quantum Information and Matter.

11 Bitcoin is periodically "minted" or created from nothing, with a method called mining, a process linked to the intrinsic functioning of the block chain. The term "bitcoin" also refers to open source software designed to implement the communication protocol and the resulting peer-to-peer network.

12 Satoshi Nakamoto is the pseudonym of the inventor of the cryptocurrency bitcoin (code: BTC or XBT).

13 David Lee Chaum (born 1955) is an American computer scientist and cryptographer. He is known as a pioneer in encryption and privacy-preserving technologies and is widely recognized as the inventor of digital money.

14 Nicholas Szabo is a computer scientist, legal scholar, and cryptographer known for his research in digital contracts and digital currency. He graduated from the University of Washington with a BA in computer science in 1989 and received his JD from George Washington University Law School.

# 4 Automation and Omination

## 4.1 THE CHANGING WORLD

We live in an era characterized by a **profound change** in the relationship between **humans and machines**. It was 2004, when **Io, robot** (I, Robot), was released in cinemas, a film inspired by **Isaac Asimov**, directed by Alex Proyas, and set in the year 2035 Chicago, a city in which coexistence with humanoid robots governed by the **three laws of robotics** (Figure 4.1), which regulated the relationship between humans and robots. The film represents a reality in which robots have now become a *household item* like many others, within everyone's reach and present in all homes.

The world of 2035 was anxiously awaiting the market launch of the brand new **NS-5**, a generation produced by US Robots, a leading company in robotics at the time. A brilliant, **Will Smith** played the detective **Del Spooner** remained famous for his distrust of the new, highly advanced robots. In fact, while everyone went crazy for these mechanical helpers, the detective, together with Dr. Susan Calvin (an expert psychologist in artificial intelligence) was called to investigate the murder of Dr. Alfred Lanning, founder of US Robots. The investigations seemed to lead to a crime committed by a robot! **"Sonny"**, another undisputed protagonist of the file, is an NS-5 considered "imperfect" because it is able to dream and feel emotions, the two characteristics that differentiate man from any machine nowadays. Several years have passed, and today it no longer seems so strange to think that a robot can feel emotions, and the famous phrase uttered by Spooner *emotions don't seem to be a very useful simulation for a robot ... I wouldn't want my toaster or the vacuum cleaners were so emotional ...*, he suggests.

In that regard it is worth mentioning, **Boston Dynamics**, a US engineering and robotics company known for the development of various robots including the **BigDog**, a quadrupedal robot designed for the US military, with funding from DARPA (Defense Advanced Research Projects Agency) and in 2013 purchased by Google. Boston Dynamics, founded in 1992 as a spin-off of MIT – Massachusetts Institute of Technology is one of the most important robotics companies in the world. One of the most interesting robots from Boston Dynamics. Figure 4.2 shows the main robots produced by Boston Dynamics.

It is clear that the perception of robots has changed in our lives and it will change more and more. **Stephen Hawking's** quote *Unless humanity redesigns itself by changing its DNA through the alteration of our genetic heritage, computer-generated robots will conquer our world* resonates even more current and even a little disturbing. In recent years, progress in robotics has greatly accelerated, thanks also to advances in other disciplines such as computer science and engineering. Today robots being networked and always connected, devices can learn at an ever

DOI: 10.1201/b22968-4

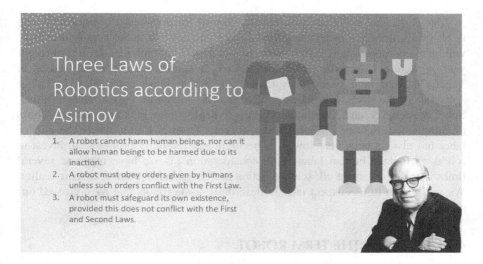

**FIGURE 4.1**    The three laws of robotics according to Asimov.

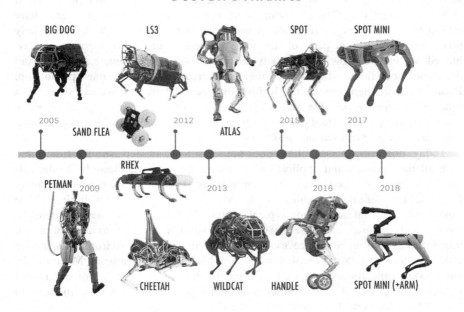

**FIGURE 4.2**    Historical evolution of Boston Dynamics robots.

faster pace from other robots. Furthermore, thanks to the development of algorithms that are able to model and make sense of unprecedented contexts, the so-called belief space. Robots are also developing an intuitive understanding. The real question is not whether machines can think, but whether humans can.

**Alec Ross**, senior advisor for innovation during the presidency of Barack Obama in his best seller *"The Industries of the Future"* argues that in the near future, there will be many more robots than humans in factories. Just think that already in recent years **Foxconn International Holdings Ltd.,** a multinational company based in Tapei, Taiwan, mainly produces under contract to other companies, including Amazon, Apple, Dell, HP, Microsoft, Motorola, Nintendo, Nokia, HMD Global, Sony, BlackBerry, and Xiaomi, is planning to deploy more than **one million robots** to complement the one million human workers. Even in China, where the cost of labor has always been very low, it is becoming increasingly profitable to invest in robots to replace human labor. One consideration (as we have repeated several times in this text) above all: science fiction and imagination often anticipate reality. We will return to this aspect later in describing some robots that have crossed our recent history.

## 4.2  BIRTH OF THE TERM ROBOT

The term **robot** was born in 1920 from the drama *RUR (Rossumovi univerzální roboti)* by the Czech **Karel Čapek**. In this drama, Dr. Rossum created *"robota"* (Czech means workers) able to work for a man so that he could free humanity from slavery. However, the epilogue was tragic, as the robots took over humanity. It should be noted that Čapek's robots were not actually robots in the sense later attributed to the term, or mechanical automata, but beings "built" by artificially producing the different parts of the body and assembling them together. Today, indeed, we can see how many robots make it possible to eliminate repetitive and dangerous operations, which is why they are increasingly used in mass production. Returning to the definition of the **Robot Institute of America**, we can state that a robot is *a reprogrammable and multifunctional manipulator designed to move materials, components, tools or specialized devices through various programmed movements to perform various tasks* (1979).

Today, however, this definition does not seem to be in step with the times, and with all the versions and applications of robots that we will describe in the following chapter. We could say that, while **Isaac Asimov's** vision was extremely futuristic, this definition is limited to the concept of robotic system, limited to the aspects of industrial automation typical of the 70s. However, the developments of robotic technologies since the 1980s have revolutionized the concept of robot, transforming it from a service tool in structured industrial environments, into a utility system and "man's collaborator" in his own environment. Most of the anthropomorphic robots that we often see in the photos or films of car manufacturing companies (Figure 4.3) are abundantly present but locked up in cages to protect humans from their movements or risky operations.

Basically, there are barriers between robots and humans, at least in the production we know today. But barriers and walls are born to be torn down. **Collaborative robotics** is born today. Robots that can insist in the same area as man and thanks to a dense network of sensors that allow for interaction with man himself. Companies like ABB have been proposing innovative solutions for some time such as **YuMi,** a two-armed collaborative robot equipped with flexible hands,

**FIGURE 4.3**  Examples of anthropomorphic robots in automotive sector.

part feeding systems, part recognition camera, and advanced control. YuMi is an example of a cobot destined to change the concept of assembly automation forever. YuMi means *you and me*, working together, side by side and without barriers.[1]

These modern "robots" are reprogrammable electromechanical systems, endowed with perception capabilities and their own intelligence, designed to perform a large number of different tasks. But let's see what was the history of robotics and how we arrived at **Cobots**.

## 4.3  HISTORY OF ROBOTICS: BETWEEN MYTHS AND LEGENDS

The idea of creating autonomous and intelligent mechanical systems is quite ancient and represents the synthesis between the dream of imitating nature and the need to build machines that are useful for life and work.

The presence of automatons, that is of artificial beings, represents a constant that uniformly crosses different times and cultures. Traces of it are found everywhere: just think of the **Golem** of Jewish folklore (clay statue created by the faithful and animated by dances and prayers) or the **Tupilak** of the legends of the Inuit, the inhabitants of Greenland (a supernatural entity created by a sorcerer and "programmed" to fight enemies). Figure 4.4 shows examples of Golem (left) and Tupilak (right).

Automatons are already present in mythology: in **Homer's** time the first creator of *machines* was Hephaestus, god of fire, who was credited with the construction of self-propelled machines and mechanical servants that moved of their own free will. Between myth and history is Daedalus, father of Icarus, to whom mythology attributes the ability to *instill* movement in the objects he created. Daedalus marks the origin of metalworking, the rules of architecture, and the first wooden statues, which, according to tradition, automatically moved their eyes, arms, and legs.

**FIGURE 4.4**    Golem (left) and Tupilak (right).

Between the fifth and fourth centuries B.C. the world of myth gradually gave way to science, and even automatons became products of man. In the Hellenistic age, intellects of the caliber of Archimedes and Heron provided a great contribution to the scientific and technological revolution: many of the principles of physics, mathematics, mechanics, and astronomy that they gave birth to are still valid today. Hero's aeolipile or **sphere of Aeolus**, for example, is an interesting tool that shows how thermal energy can be transformed into mechanical energy. It is the forerunner of steam engines; in fact, it exploits the pressure deriving from the heating of the water inside a metal sphere to generate movement.

**Arab authors** also conceived complex devices. Between 1204 and 1206 **Al-Jazari**[2] was the famous Arab scientist to whom we owe the book of knowledge of ingenious mechanisms. Today is considered the culmination of Arab mechanics, elaborated numerous projects of a *robotic* nature. **Japanese culture** has the primacy of having generated and accumulated a wealth of knowledge and made robots become *fashionable*. During the period of the so-called **Edo Era** (between 1600 and 1867) the **Karakuri Ningyo**, a sort of puppet animated by levers, wheels, and cams, was widespread in Japan. The drafting of the manuals necessary for the construction of these dolls created to amuse both the nobles and the population, still represents the foundation of today's engineering culture in Japan.

Retracing the history of robots, **Italy too** has played a leading role in the desire to be able to animate artificial creatures. In fact, around 1495 **Leonardo da Vinci** designed the **mechanical knight** automaton, the prototype of a mechanical automaton with armor whose purpose was both to enliven the festivities at the Sforzesco castle. But it is not known whether the automaton, of which the design sketches have been found, was actually built.

Instead, the one who created the first known functioning robot was **Jacques de Vaucanson**,[3] who in 1738 manufactured a flute-playing android capable of performing complex movements; another famous creation of his was **the mechanical duck,** capable of eating and defecating.

Between 1768 and 1774 **Pierre-Louis Jaquet-Droz**[4] and his son Henry built a **series of automatons** and mechanical objects, which can now be seen at the **Musée d'Histoire in Neuchatel.**

Among the historically memorable pieces, mention should also be made of the **Psaltery Player,** built in 1780 by the Germans **Roentgen and Kintzing** for Queen Marie Antoinette.

In 1769 **Wolfgang De Kempelen,** a Hungarian inventor in the service of Empress Maria Theresa of Vienna, developed the **chess player,** a mechanism ostensibly capable of playing chess automatically but actually animated via an internal linkage system by a hidden human player inside the device. Numerous contributions can be traced back to Kempelen in terms of research in robotics including the talking machine, described in the *Mechanismus der menschlichen Sprache nebst Beschreibung einer sprechenden Maschine* (1791). Such proto-automata, however, remained non-programmable and, in fact, non-autonomous mechanisms.

To approach models more similar to those of our recent history, we have to get to the 70s, when automatic machines were introduced to support industrial production. In this period, the environment in which the robot operates and the environment in which humans operate are completely separated in order to guarantee the safety of the operators. The first robots also operated in fully structured environments, i.e. where the positions of all the elements with which the robot had to interact were known in advance. Subsequent developments in robotics, such as computer vision, have made it possible to reduce the constraints imposed on the environment. The *liberation* of robots from the cages and structured environments in which they are forced into the industrial sector began in the 80s with the development of service robotics. The so-called *service robots,* such as bomb disposal robots, robots for missions in space or undersea, surgery robots (which do not replace the surgeon but enhance his skills), or assistance robots (capable of performing daily utility operations such as cleaning, washing, and serving meals – a famous example is the iCub™ project, a humanoid robot designed in the laboratories of the Italian Institute of Technology[5]), or even those designed for various types of entertainment, are gaining more and more credit and interest. A recent and extremely promising development of robotics has occurred in the field of miniaturization, which makes it possible to create very small robots capable of moving in the cavities of the human body, to carry out the so-called intraluminal surgery (these robots can be swallowed inject). The research areas of robotics are diverse. In the awareness that the fields of application of robots are numerous and constantly evolving, we will stop to analyze the industrial, social, medical, and environmental fields.

## 4.4 ROBOTS IN THE INDUSTRIAL SECTOR

Robots in the industrial sector are first in order of application, and now they are widely used in all production sectors. In fact, they make it possible to reduce costs and production times and to be able to modulate costs based on the production volumes required. In the field of industrial robotics, there are manipulators for dozens of different applications, just to name a few: painting, assembly, cleaning, inspection, welding, assembly, handling, cutting, quality control, surveying, and the number of applications seems destined to increase with the time and the autonomy achieved by these devices.

It is important to specify that the development of Industry 4.0 has favored and is favoring the development of knowledge and the diffusion of industrial robotics even among smaller companies. Mechanical robots have reached an extremely important as they allow increased efficiency in production and to develop new business paths. The mechanical structures of a robot can vary according to the automatic system under consideration. Currently, on the market and in the industrial sector there are robots that have recognized and common mechanical structures. Usually, it must be considered that there are first-level robots, second-level robots, and third-level robots:

- **First-level robots.** A first-level industrial robot is programmed to carry out and faithfully reproduce the task that has been assigned to it through specific software.
- **Second-level robots.** They are robots that have a flexible internal system capable of adapting to different working situations.
- **Third-level robots.** Third-level robots are equipped with neural networks. Thanks to these incredible components, robots are capable of making decisions with total autonomy.

In the next few years, we will be able to witness an incredible evolution of human-machine interaction thanks to the integration of artificial intelligence, deep learning, machine learning, IoT, and cyber security. We must be aware that the greatest innovations in history come from moments of great difficulty as they always have, and as demonstrated by the moments of real momentum and change in the past. We must not underestimate that the moment of crisis linked to COVID-19 could represent the best time to invest in robotics as well; we must do it with ideas capable of changing the world. The future of industrial automation will be characterized, for example, by networked **nanotech entirely based on wireless, cloud,** and **edge computing,** evolution of collaborative and mobile robotics, **Real time intelligence, Edge analytics,** and complex adaptive systems. The market requires increasingly customized products, so machines must be able to adapt to changing demand conditions. 5G will represent a further acceleration in the innovative path toward the connected factory paradigm, with, for example, 5G collaborative robot systems. The futurist **Martin Ford** in his best seller *Rise of the Robots* theorizes that the technological revolution of these years will hit harder than we are willing (or ready) to admit. And, above all, the robots that will *steal our jobs* are everyone's problem, regardless of the level of education and the type of work performed. The question is delicate. In our view, it cannot be addressed without a long-term socio-economic analysis. Surely, it is not difficult to imagine that the impact of robots on jobs will have such a strength and dimension that in the future a decrease in the classic and traditional job opportunities in the world can be foreseen. It is also true that today there are many jobs that pay little or nothing. In this sense, new technologies and robots will represent an opportunity to give more value to time and increase the quality of life of people forced to carry out strenuous work.

## 4.5   ROBOTS IN THE SOCIAL SPHERE

They identify a new application of robotics destined to be a tool for social interaction in the future (Figure 4.5). Once the safety barrier has been eliminated, through an accurate series of standards and certifications, robots can become part of **social life**. To date, the development of social robots is still limited, ranging from a few commercial examples to mostly experimental results. Robots for gaming (Sony Aibò, Mitsubishi Wakamaru, Lego Mindstorm), cinema (animatronics), and assistance to the elderly (home automation), are now widespread. **Romeo** robot is, for example, a humanoid being tested at the French company **Softbank Robotics**, designed to assist the elderly and people with reduced motor skills.[6] Another type of application is **Paro**, an interactive robot created by **Takanori Shibata**, chief researcher of the leading Japanese company Aist. Paro actually has the appearance of a baby seal born to be used in dementia therapy in the elderly.[7] Of more recent research are robots for artistic interaction, sport, and music.

The margins for the development of robots in this area are certainly enormous. The degree of diffusion in which robots will be able to be successful also depends a lot on the culture of each person. It is clear that Western and Eastern cultures are very different. Perhaps not everyone knows that even things in Japan have a soul. The objects that accompany us in our lives and in our various activities are not mere "things" to be exploited and disposed of as we please. They have their own essence. This cultural approach so different from the West means that in the East there is much more predisposition to accept the presence of robots in daily life and not be afraid of them.

**FIGURE 4.5**   Example of robot in the social sphere.

**Source: La Stampa.**

## 4.6 ROBOTS IN THE MEDICAL FIELD

The use of robotics is revolutionizing both **surgical practice** (computer-aided surgery) and **clinical practice** (physiotherapy, technological assistance). Over the last few years, various robotic systems have been designed and marketed to complete minimally invasive surgery equipment, with robotic instrumentation capable of making the intervention process natural and intuitive. The **da Vinci surgical robot** (Intuitive Surgical), for example, allows a surgeon sitting in a position close to the patient to operate as if his hands were really grippers inside the patient's body. A complex robotic system collects the movements of the surgeon's fingers and transmits them to a robotic system for minimally invasive surgery that operates on the patient's body. A stereoscopic camera system also allows the surgeon to see directly inside the patient's body. The system produces significant positive effects: it reduces recovery times, reduces complications due to post-operative infections, and limits physical stress but of course, it requires a highly specialized medical team.

There are other very interesting examples such as the **Smart Tissue Autonomous Robot** (STAR) capable of making precision sutures. In the **therapeutic field**, there are currently several neuro-motor rehabilitation systems based on computer-controlled robotic interfaces. These systems can be applied in different areas depending on the clinical pathology. For example, active tremor dampeners may be used for multiple sclerosis or Parkinson's; recording systems and motor aids can be used for post-traumatic exercises or for particular pathologies that damage motor skills. Furthermore, in complex situations such as those of an epidemic, robotics can prove to be very useful in various strategic areas.

Let us think, for example, of **clinical assistance** which involves the use of **robots** for **telemedicine** and **telepresence**. By 2045, the percentage of elderly people in Italy will rise to 32%. We are one of the longest-lived countries in the world. To support the healthcare system in the long term and guarantee quality standards, telemedicine, and enabling technologies will be the solution. Another very topical consideration is the awareness that during events such as the health emergency due to COVID-19, the use of robotic avatars in telepresence has proved to be a fundamental prevention and monitoring tool for reducing contact between doctor and patient, for infection tracing, data collection, medical analysis models. For example, Brigham and Women's Hospital in Boston used the Boston Dynamics dog spot, for hospital care to triage patients suspected of having less severe cases of the coronavirus virus.

From televisits to telemonitoring of chronically ill or prematurely discharged patients, there are many possibilities offered by technology in the healthcare sector. Definitely, telemedicine represents a great opportunity. Thinking digital means changing your mindset. There is no longer "the medical record", there is an algorithm that extracts patient data from all hospital, administration, laboratory, radiology, specialist services, reservations, pharmacy databases, and makes them available to the doctor, nurse, administrator, aggregated according to his needs. **Israel** is the country at the forefront of the use of digital tools in the health sector. Citizens who need their own general practitioner can book an appointment via the web, all reports are transmitted electronically, everything is archived, from the surgery to the hospital, up to administrative events, in real health big data. Cross-referencing this data with the patient's personal, historical, and family information allows the doctor to anticipate the diagnosis and treatment, to move from "treating" to "taking care".

In our opinion, telemedicine and teleassistance applications will be increasingly enabled by 5G, which will contribute to the reduction of public costs with consequent savings that can be reinvested in the healthcare circuit or in other contexts, to achieve greater levels of efficiency in the management of resources already in itself rather scarce. Consider, for example, the availability of beds in hospitals, the use of which can be optimized by replacing part of the hospitalization time with remote care, de-hospitalizing the patient.

In this regard, there are already realized and very interesting cases in this area. In October 2020, a Chinese medical team performed a radical cystectomy on a patient 3,000 km away using the fifth-generation 5G network connected to a Chinese-developed surgical robot. The surgical robot connected to the 5G network is seen as one of the most promising uses of this technology, which can reach patients more easily and improve healthcare in remote areas of the planet.

## 4.7   ROBOTS AND THE ENVIRONMENT

We are generally used to seeing robots built with parts in metal, plastic, and other materials. Furthermore, they are always complete with motors, servos, and other similar parts. **Jonathan Rossiter**, Professor of Robotics at the University of Bristol, has decided to focus on another type of robot. More precisely, we were inspired by **natural biological organisms** that are able to do many things that humans are not able to do: they move in gardens, in water, and in other environments where they catch insects which they then eat, etc. This type of **biodegradable robot** could be the solution to many problems, especially those concerning the many chemicals used in agriculture to guarantee ever more abundant crops and meet the ever-increasing demands of the markets. This kind of substance is partly absorbed by cultivated species, while others remain in the ground which then end up in the groundwater. In turn, these substances, in addition to polluting the water, nourish various species which, if they increase in number, create an imbalance in the ecosystem. For example, algae love nitrates, a very common chemical substance in phytotherapeutic treatments and more. Basically, the algae, feeding on nitrates, increase in number. Negative aspect if it happens in an exaggerated way, as they deprive the water of the necessary oxygen. Thus, in this case, a **robot that eats the algae**, making it harmless, would be very useful. Oil pollution represents another environmental problem caused by man and is present everywhere, even in our seas. So, biodegradable robots that feed on the pollutants produced by petroleum would be another perfect solution to curb this problem. **Rossiter,** together with his team, creates robots that feed on pollution and to create them he was inspired in particular by two organisms: the **basking shark**, which while swimming opens its mouth to collect plankton and while doing this operation digests the food and then uses the energy to move. Instead the other organism is the **little fat girl** who uses her paddle legs to move forward. Precisely in this trajectory, an animated film was released in cinemas in 2008, made by *Pixar Animation Studios* in co-production with *Walt Disney Pictures*. The protagonist of the film was the robot **WALL•E**, a small robot, the only inhabitant, in the distant future, of planet Earth, now abandoned by human beings due to excessive pollution and the continuous accumulation of waste. WALL•E's job was to clean up the planet by compacting garbage, a job he'd been diligently performing for more than 700 years.

## 4.8  COBOTS AND HUMANOIDS

In 2003, an advanced humanoid robot **Asimo** (acronym for Advanced Step in Innovative Mobility) was presented, capable of climbing stairs on its own. Unfortunately, as can be seen in a video that went viral, on reaching the third step, the poor robot trips and falls down the stairs to everyone's amazement and hilarity.[8] At the end of the decade another type of robot arrives from Boston Dynamics, **Parkour Atlas**, which instead is able to do things that only the most trained athletes in the world can think of performing, adding uncommon precision and strength.[9] Asimo, significantly improved and enhanced after its unfortunate debut, is the brainchild of Honda, born as a testimonial of the technological power of the large Japanese multinational and used on more than one occasion as Japan's representative in diplomatic visits. The highly advanced C-3PO from Star Wars is hardly a toy compared to the advanced technologies that are offered to us by robotics. And what about the fantastic **NAO**, an autonomous and programmable medium-sized humanoid robot developed by **Aldebaran Robotics**, a French technology company, with headquarters in Paris. It was presented to the public for the first time at the end of 2006. The first few times it appeared on the consumer market it seemed nothing short of science fiction. The project was launched in 2004. In August 2007, NAO replaced Aibo, Sony's robot dog, in the international **RoboCup** (Robot Soccer World Cup) competitions. RoboCup is an initiative conceived in 1993 and launched in 1997 with the aim of creating, by 2050, a team of autonomous humanoid robots capable of challenging and, possibly, beating the world champion soccer team.

Predecessors of the humanoids were some robots who became famous in the cinematographic field. Seven years after the release of Karel's robot Čapek's drama, robots make their cinema debut, in **Fritz Lang's masterpiece**[10] *Metropolis*. Few films have left such a deep imprint on the collective imagination as Metropolis, from 1927. The protagonist of the film is HEL, an automaton with female features that will become a symbol of cybernetic sensuality, the precursor of the work of the master Hajime **Sorayama**.[11] The difficulty of animating the movements of the shiny steel figure in a credible way is solved with a literary device that will be repeated in many films. HEL will take on the appearance of a real woman (Figure 4.6). We will find this solution in various films up to **2001: A Space Odyssey** by **Stanley Kubrick** where the concept of human features is overcome: the HAL 9000 supercomputer equipped with a vigorous, great artificial intelligence, capable of conversing with human beings and reproducing all human cognitive activities with speed and safety, a sort of artificial intelligence that anticipated that of today's neural networks. The imprint left by the Metropolis robot on contemporary science fiction cinema has been very incisive. Suffice it to say that the

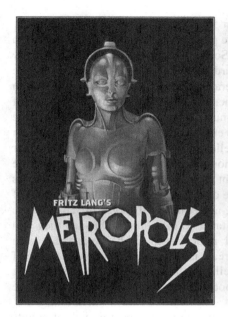 

**FIGURE 4.6**   Examples of humanoid robots (on the left HEL automaton with female features; on the right C-3PO automaton with male features).

physical structure of the famous **robot C-3PO** (D-3BO in the Italian version) and his friend **R2-D2 (C1-P8 in the Italian version), a character from the Star Wars** science fiction universe, is clearly inspired by the creature of Metropolis. A robot that has been present in all the episodes of the Star Wars saga, from the first to the last Star Wars: the rise of Skywalker, keeping alive in the collective imagination an image that comes from the 1920s.

In these years of **profound change** in production methods, cobots are proving to be able to support business growth and ensure continuity of operations. Some futurists think that there will be a near future in which robots will program themselves and in which they will increasingly have a humanoid appearance, up to the point of imagining they have a soul (which, among other things, is typical of Japanese culture). In our opinion, the impulse toward the digitization of factories due to the introduction and diffusion of the production paradigms of Industry 4.0 has given renewed importance to the use of robots, as interconnected, highly digitized tools, equipped with a digital twin, capable of improving their performance and self-learning based on the analysis of data collected in production systems. From this point of view, artificial intelligence will increasingly have to be combined with industrial robotics to make robots aware agents and able to learn from their work. Collaborative robotics, in which man and robot share the same workspace in safety, participating in the production process, today still constitutes only a niche, equal to a few percentage points of the global industrial robotics market. According to all analysts, collaborative robotics is destined to experience strong growth in the coming years, with new subjects entering the market and the big players organizing themselves to compete on this front as well. The industry needs to be research-led in the process of restructuring production lines from purely manual or purely automated to mixed production lines. In the preparatory work for future European initiatives, the **European Economic and Social Committee** (EESC) has already begun to envisage the next transition toward Industry 5.0 as characterized by the passage from coexistence to full cooperation, physical and social, between machines and people. One of the social emergencies in the industry is made up of diseases associated with stress on the musculoskeletal system. Collaborative robotics provides a natural response to this emergency since the cobot can interact directly with humans, for example, by relieving them of the task of lifting or holding heavy loads, improving the ergonomics of the workstation. Beyond the technological aspects, robotics is important from a social point of view. Robots are becoming more and more at the service of man, and we shouldn't be afraid of them. They will be increasingly accepted, as is already the case in Japan where humanoid robots are widely used for assistance to the elderly. One day they will be so widespread that we will no longer notice them, as happens today with personal computers or smartphones. The dream is to arrive at machines capable of developing feelings. Machines are so intelligent that in the future they will be able to choose, out of sympathy, which robot or human being to relate to. Just like it happens to each of us in daily life.

It is thought that in the next years, there will be a significant growth of **cobots** equal to the number of traditional industrial ones. This provides us with the possibility of so-called **back rescoring** as they significantly reduce labor costs. Service robotics is increasingly present in our lives. Starting from this fact, Bill Gates provocatively launched the idea of taxing robots a few years ago. Even if in nature we find living beings that are able to make movements and articulations that today it seems that traditional robots will never be able to do, it probably won't be long before we can see robots capable of breathing, feeding, and dying.

It is important to remember that 2017 was the year in which many humanoid robots made headlines thanks to their participation in various TV shows. Robotics companies producing humanoids are on the rise. Currently, the United States is leading the robotics sector, but according to forecasts, China and Japan will gain ground in the coming years. Humanoids will also be used more and more in medicine, commerce, logistics, education, and other sectors as well. While the research continues, there are some models that have already been launched and have achieved great media success. For example, **Pepper**, the robot made by Softbank Robotics who works as a receptionist, waiter, or clerk in Japan. About one meter tall, Pepper (Figure 4.7) is a robot capable of interacting with human beings, following them with his gaze during their movements, and reacting to stimuli that come from the outside, such as questions or caresses.[12]

Precisely, because of its characteristics, Pepper has become a protagonist in times of COVID-19 as a support in some high schools to screen and measure the temperature of all the people who access the facilities.

While **Reem, the first robot policeman,** took service, while at the Asian Oktoberfest in Quingdao in eastern China, customers were served by a humanoid.

The most famous humanoid is certainly **Sophia**, the "woman" robot with incredible human features who, in addition to having an upright posture, can also

**FIGURE 4.7**  Pepper Robot.

**FIGURE 4.8**  Images of Sophia and her creator David Hanson.

converse naturally (Figure 4.8). Sophia was created by **Hanson Robotics**, a Hong Kong company founded by sculptor and special effects expert David Hanson, who became famous for the humanoids they created. Sophia is the company's most famous robot. A humanoid who even managed to obtain citizenship of Saudi Arabia.

Another famous humanoid is **Geminoid**, created by the Japanese scientist **Hiroshi Ishiguro,** Director of the Artificial Intelligence Laboratory of Osaka University. Geminoid is the exact copy of itself. In other words, Geminoid robot is an android physically identical to a real man, a "clone" capable of deceiving even the most attentive observer. For years Hiroshi Ishiguro, together with his research team, has been working to make the interaction between man and robot as natural and credible as possible. To do this, he created a "family" of humanoid robots. Figure 4.9 shows examples of Geminoid M (identical to the scientist) and Geminoid F.

When we talk about the replacement of humans by humanoid robots, the thought quickly falls on **PromoBot**, a Russian startup that operates in 35 countries with various partnerships. The rather shocking novelty was announced by Aleksei

**FIGURE 4.9**   Examples of Geminoid robots.

Iuzhakov, chairman of the board: "Robo-C is a human android that simulates the appearance of a person and integrates into business processes". Robo-C is able to move his eyes, eyebrows, lips, and muscles, participate in conversation, and answer questions. Incredible realism of the patented artificial skin and capable of reproducing around 600 facial expressions. The robot is equipped with 18 mobile points in the face and is enriched with 100,000 artificially created voice modules. It can be ordered à la carte and also includes the production of celebrity robots such as **Arnold Schwarzenegger**. According to the Russian engineers, these robots can be used in fairs, airports, museums, schools, hospitality, and commerce.

## 4.9   ROBOTS, SOUL, AND ANDRORITHMS

One of the most debated aspects is certainly the one related to the similarity between robots and humans. *Will robots ever have a soul? Will they be able to emulate humans?* At the moment it is not easy to provide an answer to these questions. Science will have to make significant leaps forward in the field of artificial intelligence for this scenario to become a reality. One thing is certain: many researchers and scientists are working to pursue this goal. This is precisely Ishiguro's goal, that is, to give the Geminoid robots a soul, a *thought and a conscience* of their own. Only at that point can the emulation process be considered completed. We still don't know if he will succeed or if those who come after him will succeed. But the journey has already begun.

From our point of view, the real goal is not so much to design robots that emulate humans but rather to understand what intelligence is, whether it is something limited to the mind or whether it also concerns the body. The concept is well explained by the futurist **Gerd Leonhard,** who, in his new book *Technology Vs Humanity*, introduces the neologism *androrithms* which he created to describe what really matters to us: human "rhythms", not machine rhythms, i.e. algorithms . A super-computer can win at chess or GO, but it can't currently talk to a two-year-old. A person who meets us in a corridor needs an average of 1.4 seconds to get a basic understanding of us, even without speaking. A computer still doesn't understand our values and feelings even after processing our browsing and social networking histories of the last seven years (about 200 million data). Digital transformation, exponential progress, the future of the Internet and of work, artificial intelligence are challenges and as such should be "exploited" and not "personified". From this point of view, many scientists and scholars such as **Jeffrey Schnapp,** director of the metaLAB at Harvard and pioneer of the so-called digital humanities, invite us to consider the limits of technology and in particular of robots. Schnapp states "Don't think of robots as equal to us but next to us". Other scholars, including **Raymond C. Kurzweil**, an American inventor and computer scientist, argues that the consequences of perfecting AI will lead to an inescapable reversal that will lead humanity to hybridize with machines to achieve transfer from our biological body. He argues that we will soon be able to transfer our consciousness to a computer, thus providing us with a form of immortality. Biological man would then be only a stage in evolution. The following phase would mark the advent of the "bionic man", a robotic system commanded by a human consciousness freed from its biological nature and registered on a machine. Kurzweil places the realization of this critical moment, the **Technological Singularity**, in **2045**. This hypothetical moment would represent the stage in which AI will surpass human intelligence. Many questions and doubts remain open. It is certainly premature to want to provide a definitive answer to the question of whether humanity will hybridize with machines to achieve transfer from our biological body. We are only at the beginning of an era of major changes, and we just have to open our minds and be ready to welcome them.

## 4.10 THE UPCOMING CHALLENGES OF ROBOTICS

In light of the previous considerations, one might wonder what the next challenges of robotics will be and how the development of artificial intelligence will be integrated and integrable with these systems. In this regard, robotic technology is destined to influence every aspect of people's work and life, offering the possibility of transforming them positively, increasing levels of efficiency and safety, and providing new services and answers to the needs of a population growing in middle age. Robotics is set to become the driving technology behind a whole new generation of autonomous devices that, through their learning capabilities, can interact with the world around them. Thus, they provide the missing link between the digital and physical worlds we live in. In this sense, we explain the neologism **Interaction Technologies** (InterAction Technologies, IAT) introduced to explain how robotics and intelligent machines represent the future of those **Information Technologies** (ICT) which today stop at the level of data collection and processing. However, robotics will unfold their full potential only when they can be used to physically intervene on the environment and on people, to modify the former and assist the latter with the ability to perceive and act in the physical world in real time. Robotics is already the engine of competitiveness and flexibility in large-scale manufacturing industries. Without robotics, many of the successful manufacturing industries would not be able to compete nor will they be able to re-attract those productions that have been delocalized. While in the large-scale manufacturing sector robotics is already a fundamental tool for productivity, it is thanks to the development of new collaborative robotics technologies that innovation is becoming increasingly relevant even for smaller industries. Service robotics will show even more disruptive effects on the competitiveness of non-manufacturing industries such as agriculture, transportation, health care, security, and public services or utilities. Growth in these areas over the next decade will be explosive. From what is currently a relatively small starting point, forecasts indicate that service robots used in non-manufacturing areas could soon become the sector of robotics with the highest added value.

Surely, COVID-19 has been a strong "technological accelerator". Things that until yesterday seemed impossible to implement have now become indispensable. In this context, robotics naturally places itself at the intersection of the transitions that advanced societies are preparing to face, both those motivated by a heightened awareness of new economic, natural, and social problems. At first sight, the interweaving of these transitions appears as an irresolvable puzzle. Really, there are so many variables and constraints present, and each transition cannot be tackled independently of the others, due to the strong interdependencies that exist. We therefore need solutions that can address these transitions in a systemic way, creating physical, economic, and environmental well-being for society as a whole in a democratic and supportive way. Robotics lends itself to being one of these solutions because its enabling technologies can be combined to create devices for the most diverse applications, maintaining the same fundamental characteristics and developing vertical supply chains that can optimize development times and market entry. The area in which intelligent robotic systems will perhaps have the greatest impact is that of the economic transition between the current model of economic development and a new sustainable model, respectful of the environment and people, a true social innovation catalyzed by robotics. Contrary to what one might think, however, the true value of robots will not only be in making production processes more efficient and in the creation of factories where workers and robots can collaborate in a safe way, but also in the recovery of activities and professions. For example, a key sector for the Italian economy is that of fashion, the production of which is however made, in most cases, in countries with very low production costs, increasingly distant from Italy. The reduction, if not the elimination, of long transport journeys puts this production model out of the market and therefore requires innovative solutions. This is the problem of reshoring, the return to advanced countries of production activities transferred to less developed countries, thanks to the use of robotic technologies that once again make these productions economically sustainable. For example, the production of clothing and shoes has largely been outsourced in the pursuit of ever lower production costs. A robotic factory with devices capable of producing clothing automatically and at low cost would make it possible to recover production activities that have disappeared in Italy for years. This recovery would not be done to the detriment of workers but rather would allow the creation of a new class of operators and technicians of robotic production, able to program and manage processes and transfer the knowledge necessary for production to the new intelligent robots. This new generation of collaborative robots will be able to learn complex actions such as assembling and sewing a piece of clothing or building a shoe. It will also make it possible to help many artisans who, as they get older, are no longer able to carry out the more physically demanding parts of their profession, nor can they find young people in the workshop who can help them. Collaborative and intelligent robotics will be a valid aid to a craftsman in carrying out his work and an attraction for young people who will no longer feel tied to a single production specialization, the one made in the craftsman's workshop, but will become robotic technicians able to work in different professions, in small and large companies.

## NOTES

1 A video of YuMi is visible at https://new.abb.com/products/robotics/it/robot-industriali/yumi.

2 Al-Jazari (Jazīra, 1136–1206) was an Arab mathematician, inventor, and mechanical engineer, the most important exponent of the Islamic tradition of technology. He was the author of the Compendium treatise on the theory and practice of mechanical arts where he described 50 mechanical devices (automata) with the instructions for building them.

3 Jacques de Vaucanson (Grenoble, February 24, 1709–Paris, November 21, 1782) was a French inventor and artist, famous for the invention and construction of numerous complex automatons.

4 Pierre Jaquet-Droz (July 28, 1721, La Chaux-de-Fonds–November 28, 1790, Bienne) was a Swiss watchmaker and Swiss inventor of the late 18th century. He lived in Paris, London, and Geneva. To promote his watchmaking business, he designed and built animated dolls, known as automata.

5 Interview with the robot iCub (https://youtu.be/mQCkYeJ7UJc).

6 Romeo Robot Project (https://youtu.be/9SpekYGi-9w).

7 PARO Robot Project (https://youtu.be/2ZUn9qtG8ow).

8 ASIMO robots (https://youtu.be/VTlV0Y5yAww).

9 Parkour Atlas (https://youtu.be/LikxFZZO2sk).

10 Fritz Lang, pseudonym of Friedrich Christian Anton Lang (Vienna, December 5, 1890–Beverly Hills, August 2, 1976), was an Austrian-born American director, screenwriter, and writer.

11 Hajime Sorayama (空山基Sorayama Hajime?; born February 22, 1947 in Imabari) is a Japanese artist and illustrator. Mostly known for its signature robot women. Between 2000 and 2012 he collaborated with Sony on the design of the Aibo prototype, obtaining Japan's Grand Prize.

12 Interview with PEPPER (https://youtu.be/zJHyaD1psMc).

# 5 Digitization in the National and International Paradigm

## 5.1 DIGITIZATION OF COMPANIES: NATIONAL AND GLOBAL PERSPECTIVE

The dramatic spread of the COVID-19 pandemic has generated socio-economic impacts such as outlining a scenario in which uncertainty becomes the distinguishing feature on a global level. The socio-economic and geo-political revolution generated by the pandemic has laid the foundations for the reconstruction of a new world order based on cooperation and collaboration between institutions and companies on a global level. Even production chains, which have shown all their fragility in this period, are destined to change, and new models will be evaluated. In fact, **relocation** has become a **boomerang**. Difficulty in finding goods and skipped production schedules. Our companies are organized in very long supply chains (due to the way our very B2B-oriented fabric is made up) and this global slowdown has interrupted the production chains. We will probably see a **reshoring** for strategic productions, but also for less obvious ones. From this point of view and in this scenario, it is clear that 4.0 technologies can offer the opportunity. Businesses should invest as much as possible in new digital technologies. On this, we all agree. The problem, as always, arises in the concrete, that is, when entrepreneurs find themselves deciding whether or not to invest in a specific digitization project, for example, in the cloud.

It is clear that the digitization of companies represents an opportunity for economic growth. It is promoting a profound global transformation, revolutionizing the way businesses operate and interact on a worldwide scale. Companies across all industries are recognizing the immense potential of digitization in enhancing efficiency, expanding reach, and unlocking new growth opportunities. From large multinational corporations to small startups, organizations are leveraging digital technologies to streamline their processes, optimize supply chains, and improve customer experiences. This global shift towards digitization has significantly impacted the global economy, fostering innovation and reshaping traditional business models. Moreover, the digitization of companies has transcended geographical boundaries, facilitating cross-border collaborations and enabling companies to tap into new markets. With increased connectivity and the rise of e-commerce, companies can now engage with customers from all corners of the world, breaking down barriers to entry and creating a level playing field for businesses of all sizes. Additionally, the digitization of companies has generated vast amounts of data,

DOI: 10.1201/b22968-5

leading to the emergence of data-driven decision-making and advanced analytics. This wealth of information enables businesses to gain valuable insights into consumer behavior, market trends, and operational efficiencies, empowering them to make informed strategic choices. However, the global perspective on digitization also entails challenges and considerations. The digital divide persists, with disparities in internet access and technological infrastructure across different regions. Bridging this gap is crucial to ensure equal participation in the digital economy and prevent further inequalities. Moreover, as companies become increasingly digitized, concerns around cybersecurity and data privacy are paramount. Organizations must invest in robust security measures and adhere to stringent data protection regulations to safeguard sensitive information. Lastly, the digitization of companies has implications for the global workforce as automation and artificial intelligence reshape job roles and demand new skill sets. Governments and educational institutions must focus on upskilling and reskilling programs to equip individuals with the necessary competencies for the digital era. In conclusion, the digitization of companies has fundamentally transformed the global business landscape, offering unparalleled opportunities for growth, innovation, and international collaboration. However, it also necessitates addressing challenges such as the digital divide, cybersecurity, and workforce transformation to ensure a sustainable and inclusive digital future.

## 5.2 SOCIAL INEQUALITY, AUTHORITARIANISM, AND THE DIGITAL DIVIDE

Social inequality is one of the main factors of the digital divide, and at the same time, the digital divide is one of the increasingly significant factors in the development and increase of social inequality, all if it is combined with territorial inequality, which favors the impoverishment of growth and autonomy and therefore social inequality.

The rise of digital technologies has the potential to bridge gaps and create opportunities, but it also has the capacity to deepen existing disparities and concentrate power in the hands of a few. Social inequality, both within and between countries, is exacerbated by the digital divide. While access to the internet and digital tools has become a necessity for participation in the modern economy and society, a significant portion of the global population still lacks access to reliable internet connections and the necessary resources to fully engage in the digital realm.

Therefore, although digital technologies have the potential to reduce the digital divide, it is necessary that the digital divide be reduced.

The accessibility of digital technologies can be improved by adopting policies that make devices and services cheaper and more accessible to all.

Furthermore, it is important to ensure digital literacy and skills training necessary to use digital technologies effectively. Digital technologies can also facilitate access to essential services such as education, healthcare, and financial services.

In our opinion, collaboration between governments, businesses, civil society organizations, and international bodies is key to addressing the digital divide.

Another no less important problem concerns digital authoritarianism. There are countries like China where digital authoritarianism is particularly felt. However, it is increasingly evident that this phenomenon cannot be considered limited to China alone, but, in different forms, affects all developed and developing countries. We explored the topic with **Maria Umar**, a young Pakistani entrepreneur and current President of the **Women's Digital League**, in an interview with us.

### Founder and President, Women's Digital League

Maria is the founder and president of Women's Digital League (WDL) –an online platform that provides digital training and work to Pakistani women. She has been working in the online digital outsourcing sphere for more than five years. Project Artemis/Goldman Sachs 10,000 Women Program was the turning point in her career when she was coached and mentored by top business professors and talent from Silicon Valley. Getting to the finale of GIST's "I Dare" business plan compe-tition was another huge encouragement. Google Pakistan profiled WDL in its online campaign, showcasing innovation in the use of technology. Maria was nominated as a Thought Leader by Ashoka Changemakers. WDL also won the Early Stage Award in Changemakers' "Women Powering Work" competition. Local and International media have featured her as an innovative leader. These publications include Mashable, Forbes, Virgin, Ashoka, and Dawn, to name a few. Maria is also actively working to encourage girls to opt for STEM at an early age through the Technovation Challenge.

*If someone had asked me, even in 2007, if the Internet could potentially become a place where I could earn a living, I would probably have been surprised. But then came 2008. Fired from a teaching job for daring to become pregnant, I was in a bad place. And then I had post-partum depression. The fateful moment came when I typed work online into my new best friend, Google, and found out about oDesk, Elance and Freelancer. Since then, I – someone who could not go out shopping without my mother and a male chaperone after 5 p.m. – have traveled to different parts of the world alone, founded an internationally recognized company and won awards for my work. Access to digital information platforms is what made it possible for me to realize my potential, earn a living and be independent. Now I am empowering all my sisters who are looking for better avenues.*

**Dear Maria,**

**could you tell us your experience and your point of view on digitization?**

**#1. Digitization and innovation are two processes that guide today's companies. In your opinion, what are the drivers and obstacles to digital transformation?**

Drivers are literacy in software and digital security. People have become more aware of the need for tech education and to make it a seamless part of our everyday education from early school education to the most basic jobs. If we could add a visible layer of protection, safety and privacy we would see even more growth in terms of digital transformation. In Pakistan, for example, we see gross human right violation with government and other agencies spying on citizens especially journalists, left wing political leaders and civil society members. Freedom of thought and speech, that we thought was finally available to all through the Internet, is turning out to be a sham. Also, volatile speech and campaigns against individuals fueled by same culprits turn into witch-hunts discouraging many from speaking their truth. Stronger Internet Freedom Laws need to be not only introduced but also implemented. Harassment is another issue many women in my country face. Every time a woman gets stronger and her voice reaches loud and strong she is attacked by trolls and openly harassed on the Internet and also threatened in real life.

**#2. Do you think the covid pandemic will be a strong technological accelerator?**

It has been a strong "technological accelerator". We have seen the education industry, for example, greatly impacted and adapting tech. However, Net fatigue is also very real. Now that work form home has become the norm somewhat, people are not respecting that work-life line. Bosses and supervisors are constantly engaging employees at all times of the day and people have no option but to respond because employment has become such a privilege. Although it has been a blessing in recent times yet human interaction has emerged as the clear victor in the debate between WFH and actual conventional workplaces.

**#3. Which skills are important in a digital environment? Is your company investing in new business models?**

My company has always been based around the ad hoc WFH model so not much changes there. However, a shift is seen in many businesses especially small home-based ones generally run by women. They are now turning to digital advertising and making their ecommerce websites. Many of these women were not tech savvy but are now using Zoom quite a bit. Effectively using social media while keeping safe has always been very important. In addition. operating ecommerce websites is emerging as a very important skill. Of course I am speaking from the perspective of women-led small businesses. Also, financial transactions using online banking is one more thing women have generally hesitated form but now they are forced into exploring the options.

## 5.3 DIGITAL ECONOMY AND NEW GLOBAL OPPORTUNITIES

In the last three decades, the intensification of free trade has led to a rapid growth in world trade in goods, especially in exports. Economic surveys have highlighted the significant benefits for developing countries. In this global market, Europe continues to be the most attractive continent in terms of consumption, always closely followed by the North American market and constantly growing by the Chinese one. Internationalized companies now see the **European market** as a **domestic market** while overseas destinations are the most sought after even if very complex, especially for B2C producers. Different dynamics of positioning of goods, defense of one's brand, logistical difficulties, legal protection, and customs taxes are just some of the characteristics that often limit the approach to promising but very competitive markets.

In this context, in our opinion, digitization can act as a gateway to new and stimulating opportunities for **emerging countries** and for **African countries**. A swift and well-structured transition to a digital economy would be needed. This would increase both productivity and the creation of added value in the various stages of production, as well as the further expansion of extra- and intra-African trade, the latter above all by virtue of the new African continental free trade area. Innovations such as **eBay** and **Paypal** play a strategic role in the economic development of countries characterized by weak financial systems. Such innovations have had a significant impact in creating codified markets that have the potential to reach the world's most isolated communities and connect emerging markets to the global economy.

Often, the **black market** dominates the economy of the poorest countries in the world such as Congo and Rwanda. Furthermore, the constant fighting devastates these territories and displaces thousands of people. However, some promising signs of a "new" economy are visible in Rwanda. In fact, the African market entered a phase of very rapid development in the 2000s: between 2004 and 2007, the expansion of mobile telephony in Africa was three times higher than the world average. The operators who introduced mobile telephony in Africa in the 2000s have chosen business models suited to the poorest segments of the population. Mobile phones designed to be very low cost have been marketed, and prepaid cards have been introduced in extremely low denominations. The diffusion of mobile telephony also often represents the only way to maintain contact between the various members of families displaced in the various territories. For this reason, alongside a level of extreme poverty, we are witnessing an apparent paradox which sees the diffusion rate of mobile telephony among the highest in the world. According to KT Rwanda Network, the company in charge of supplying the network with 4 G Long Term Evolution (LTE) technology has achieved coverage equal to 95% of the country. In 2019, **Rwanda launched the first fully African-made smartphone.** According to the President of Rwanda, **Paul Kagame**, in this way, the country has laid the foundations for becoming a modern hub of **technological innovation** and a pole of attraction for the entire Central African region. What is true for Rwanda is true for the rest of sub-Saharan Africa. It is clear that this project encompasses great prospects since the diffusion of mobile telephony represents opportunities not only for communication, but gives **a central role to mobile money and digital payments**, considered fundamental prerequisites for the expansion plan of African countries. In this regard, it should be remembered that the **M-PESA** program was introduced in Kenya, a money transfer service between users of the cell phone service born in 2007 (when the average user in Italy was still very little familiar with this type of service) on the mobile network of Safaricom, a subsidiary of Vodafone, to allow microfinance institutions to easily send and receive money from lenders. In sub-Saharan Africa, as many as three million jobs have been created thanks to **Fintech.** The digitization of financial services is increasing its turnover in a dizzying way: 40 billion dollars in the next two years, reaching 150 in 2022. New financial technologies are significantly contributing to reducing the cost of transactions and simplifying the ease of transferring money. Among other important effects of this phenomenon is the reduction of **corruption**.

## 5.4   IMPACT OF 5 G ON THE GLOBAL ECONOMY

In light of increasing digitalization, the use of data to generate value based on artificial intelligence, and the resulting requirements for manufacturing industries. For the next few years, the real and virtual worlds will continue to merge, and physical value chains will be increasingly integrated into digital value chains. Product life cycles will be integrated from idea, design, and engineering to manufacturing over production to after-sales and on-site customer service. In this scenario, the impact of **5 G** will be **disruptive** – Experts from **Frost&Sullivan** predict that by 2050, more than 80% of the population of developed countries will live in cities (although it will be necessary to see these analyses how and how much they will undergo changes as a result of the pandemic, as reported in the previous chapters). In developing countries, this figure is expected to exceed 60%. The creation of smart cities allows for a smooth transition towards urbanization and technological advances will help governments to optimize resources to provide the maximum value to the population, understood both in financial value and in terms of saving time or improving the quality of life. From urban mobility to safety, from government to health, from environmental monitoring to transport, up to the tourist offer and entertainment, tens of billions of devices and sensors are applied to things and people, with very high-performance connections that will generate an ever-increasing number of data, accompanying the evolution of the digital society over the next 20 years. This is the summary of the impact of 5 G. From an Industry 4.0 perspective, applications have in fact been launched in the field of robotics and industrial automation. According to **Gartner** estimates, 5 G is an emerging technology under development with fragmented global coverage and explored use cases, whose potential and impacts are enormous (Figure 5.1). Gartner's forecast itself places this technology among those with a high impact on business over the next two to five years.

According to analyses, the boom in 5 G infrastructure will offset, at least in part, the reduction in spending on previous generation networks: 4 G and LTE networks will drop by –20.8%, 3 G by –37.1%, and the 2 G by –40.8%.

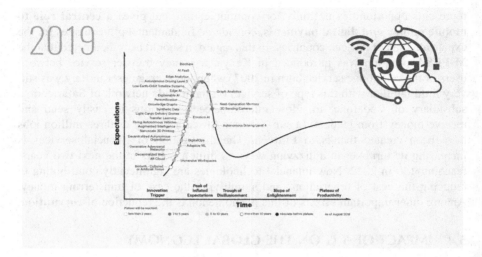

**FIGURE 5.1**   Gartner hype cycle 2019.

5 G represents a strategic opportunity for our country. However, investments must be made to improve connectivity. In some areas of the country, we are not served by fiber and the connection is bad. In order to integrate artificial intelligence processes, the right infrastructure is needed. 5 G will be the next revolution, but every single Italian municipality has its own idea about it, and there are many obstacles to its diffusion. AI is an evolution of the first phase of digitization, and with machine learning, many production processes and interactions with customers will change radically. **We have to be ready**. The competitiveness of our businesses will pass from here. We can say with certainty that the digitization process with 5 G will undergo an unprecedented acceleration, and the economic, productive, and social systems will increasingly interact with each other. In addition to the resources that fall within the "system" following the reduction of social costs, the positive impacts created at the production system level thanks to 5 G-enabled applications must be considered.

## 5.5  THE GLOBALIZATION AND DIGITIZATION: THE MAIN HIGHLIGHTS

In the last 12 months, we have seen some of the most profound changes in digital behavior, and sometimes with unexpected evolutions, that we have seen over the years, even with respect to the pandemic period. According to the Digital Report 2023,

- The number of Internet users is growing.
- The time dedicated to some activities that people carry out online is reduced.
- People's preferences in terms of social platforms evolve.
- The devices people use to access digital content and services are changing.

- Habits and behaviors when it comes to doing research online are also changing.
- Digital advertising is on the rise.

It is important to underline that the data show that:

- The world population exceeded 8 billion on November 15, 2022, and reached 8.01 billion at the beginning of 2023. Just over 57% of the world's population lives in urban settings.
- About 5.44 billion people use mobile phones, equal to 68% of the world's population. Unique mobile users increased by just over 3% last year, with 168 million new users in the last 12 months.
- There are 5.16 billion internet users; 64.4% of the world's population is now online. Total global Internet users have increased by 1.9% in the past 12 months, but some data reporting delays mean that actual growth is likely to be higher than this figure suggests.
- There are 4.76 billion social media users worldwide, equivalent to just under 60% of the world's population. Growth has slowed in recent months, with 137 million new users, equating to 3% annual growth.

Figure 5.2 shows an overview of the use of connection devices and digital growth.

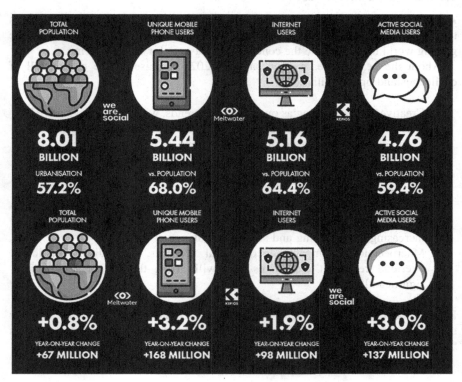

**FIGURE 5.2** Overview of the use of connection devices and digital growth.

Importantly, Internet users have grown by around 98 million in the past 12 months, an annualized growth of just under 2%, slower than the growth rates we saw in the 10s. Predictable slowdown, given that over 6 out of 10 people are connected. It is expected that 2 out of 3 will be online by the end of 2023.

The most interesting metric in this year's report is the amount of time we spend online, which is down nearly 5% year-over-year. According to GWI[1] (GlobalWebIndex), the typical user reduced the time they spend using the Internet by 20 minutes per day, compared to 12 months ago. Time spent connected dropped to 6 hours and 37 minutes a day.

Data are very close to the daily average for the third quarter of 2019, just before the COVID-19 pandemic impacted our digital (and non-digital) behaviors.

The recent easing of "zero COVID" policies in China could lead local users to spend more time in the "real world" in the coming weeks, potentially leading them to spend less time online, and given that China accounts for more than one in five users of Internet of the world, any changes in their online behaviors are likely to have a visible impact on global averages as well. Crucially, the analysis indicates that people are taking a more thoughtful and focused approach to their online activities.

In other words, people prioritize quality over quantity.

### What about the Metaverse?

All virtual worlds – metaverses – are trying to gain users, but the latest data underlines that the appeal of these worlds has not yet reached significant peaks, leaving the gaming world to consolidate the concept of metaverse for the time being.

About the near future, we can say that 2023 promises to be a year full of challenges and opportunities.

- Creative AI is emerging as one of the key trends, with tools like ChatGPT, Dall-E, Midjourney, Stablediffusion, and Synthesia representing just the beginning of a new wave of AI that can create content. We expect significant performance improvements and increased adoption of real-life applications, accompanied by a growing debate about legality, ethics, and the potential for misuse.
- In the field of marketing, we expect a greater focus on adding value to people's lives, with consequences for advertisers who will have to adapt to increasingly demanding and selective audiences.
- Macroeconomic impacts, geopolitical tensions, and disinformation overload represent challenges that will affect the digital landscape significantly.

Finally, social platforms will continue to expand their role in people's daily lives, but there is growing interest in reflecting on generic terms such as "social media".

## NOTE

1 GWI is an audience research company founded by Tom Smith in 2009. The company provides audience insight to publishers, media agencies, and marketers around the world. GWI profiles consumers across 48+ countries with a panel representing over 2.7 billion digital consumers, making insights available through a subscription-based platform.

# 6 Re-Skilling, Up-Skilling, and Human-Skilling

## 6.1 ENHANCEMENT OF THE HUMAN FACTOR AND "HYBRID" PROFILES

A fundamental prerequisite for dealing with the profound transformation that is pervading our society concerns digital skills. The new global geopolitical and economic arrangements are forcing a serious reflection on the strategic priorities of the coming years. In this context, the transformation of factories and businesses, with the introduction of new professional figures with new digital skills, plays a key role in defining the company's strategy. Ensure a work environment that maximizes the creative and distinctive intelligence of human activity. Training is a fundamental requirement for the development of **competitiveness**. The Fourth Industrial Revolution is changing work and the demand for work at great speed, with radical and intragenerational transformations that require basic skills, a predisposition to change, and increasingly articulated and complex skills. The Fourth Industrial Revolution also placed a tipping point between the pre-digital world and today's. Technology in its proactive evolution stimulates to innovate production processes and at the same time obliges an approach of **continuous training** with an inviolable common denominator the enhancement of Human Capital.

Digital Transformation, understood as the digitization of processes and evolution of the production chain, has also begun to contaminate SMEs as well if sometimes not consciously. This path, however, requires more and more professional profiles with scientific, technological, engineering, and mathematical skills (the so-called **STEM** professions). The change underway inevitably affects all companies in their business, organizational, and production models: technical skills, however necessary, are no longer sufficient, just as the simple investment in technology, which "does not guarantee the achievement and maintenance of a competitive advantage; the workforce must also be suitably informed and trained". The main recruiting companies, as mentioned in the 2020 Report *Randstad, Industry 4.0 mismatching and reskilling,* indeed agree on the need to develop new professional profiles capable of managing the growing technological complexity, but also on the urgency of retraining existing profiles under the banner of greater autonomy and the strengthening of transversal skills or soft skills.[1] The basic orientation is that of **lifelong learning**, i.e. continuous learning, both from the point of view of workers, as the new necessary skills appear much faster than the old ones take to disappear, and from that of companies, who will have to select their collaborators not so much and not only on the basis of the knowledge acquired, but by evaluating their ability to learn new skills. In this sense, we point out the worthy initiatives of some companies that are committed to training their talents internally, building training courses tailored to their needs; however, we

DOI: 10.1201/b22968-6

believe above all in the importance of a systemic collaboration between the business world and the world of training (high schools, universities, etc.) in order to concentrate resources and create profiles in line with the market right from the moment of exiting institutional educational paths. The growing diffusion and valorization of industrial doctorates represents, from our point of view, a valid example of innovation as the result of a synthesis between academic research and business strategy.

Alongside continuous professional development, we consider **transversal skills** or **soft skills to be strategic**. The most requested professions in recent years are the so-called **hybrid profiles**, i.e. roles characterized by a strong technical competence, but also strong relational skills such as the ability to work in a team, problem solving, and communication; indeed, double-digit growth is expected in the demand for these figures, characterized by strong judgment and analytical skills.[2] Indeed, the reality in which we live is now invaded by technology in every aspect. If then the technology is intended for less and less specialized users and the context of use is ever wider, it can be understood that even the more technical professions, such as the programmer, more and more transversal skills will be required to develop empathy towards the end user and know how to coordinate within an interdisciplinary design team. Added to all this are **digital skills**, i.e. the ability to interact familiarly and critically with ICT, which have become essential during the lockdown period even for professions in which they were not crucial before. These skills are enabling, for example, for effective **smart working**, which can establish itself as a rewarding and stable way of working even beyond the health emergency and allows, for example, to draw on a pool of talents less constrained by geographical location. However, to move on to considering smart working as an opportunity rather than just a forced reaction and to fully implement a flexible and goal-oriented way of working rather than the hours worked, a purely cultural change is necessary as well as relating to technological infrastructure: the entire organization must therefore undertake a participatory process of change aimed at consolidating a relationship with collaborators which, based on trust and responsibility, allows profitable flexibility for both parties.

## 6.2 JOBS OF THE FUTURE: THE ECONOMY WILL BE EXPERIENTIAL

What are the professions most in demand by companies in the future? Which schools should we direct our young people to to facilitate their job search?

We are deeply convinced that from now on, educators must keep in mind the development of young people as subjects able to contribute to the whole of society. Training, in our opinion, will have to focus primarily on technology. In fact, digital tools make it possible to develop content for remote learning (e.g. delivery of blended training courses), thus facilitating access to education for students from all over the world. Furthermore, with reference to the concept of resilience, it is necessary to equip the business model with flexible tools on the revenue side and on the cost structure, capable of absorbing significant variations and high volatility rates on both the demand and supply side. Furthermore, constant learning (continuous learning) must represent a crucial objective in all workers (the principle of lifelong learning must characterize the life of all those engaged in work activities). Finally, the need to develop entrepreneurial skills lies in the fact that it is necessary to equip the organization with strategic leadership models capable of ensuring business continuity. Personnel training is an inescapable pillar of the company's organizational evolution. It is in fact a concrete need comparable (for the most far-sighted) to the need to have new production machinery when market demand is expected to grow. The changes that will most impact businesses in the next five to ten years show the need for specialists capable of optimizing business processes and controlling their costs, as well as people skilled in development and innovation, in computerized industry, in research new markets, and in adapting to customer needs. More specifically, by digital professions, we must mean professionals with a common denominator: one works on the web, on software, on networks and applications, mobile applications, digital data and analytics, artificial intelligence and automation, connectivity, communication, training, and management in the digitized economy. According to the latest OECD survey, the most requested professions in the next five years will be experts in cyber security, blockchain, and data scientists. These subjects, according to this study, attract only 25% of students in Italy against a European average of 37%, despite the fact that the labor market is increasingly hungry for professionals specializing in digital. Also widening the gap is the lack of a training offer that is truly in step with the demands of businesses. Companies will increasingly be looking for highly specialized figures. A series of works related to new technologies and developing network infrastructures, starting with 5 G, which will increasingly enable the Internet of Things. In this regard, the *Bureau of Labor Statistics* also claims that in the near future, there will be a huge demand for people with information security skills. We must expect that the development of digitization will also have its negative sides, as **Martin Ford** reminds us in his book. The first will be the almost **total disappearance** of some **professions**, as it is possible to detect from the BCC study[3] in which it is hypothesized that, according to a precise algorithm put in place by **Michael Osborne** and **Carl Frey** of the University of Oxford, even the years that will be necessary for a profession to disappear completely from the labor market. The professional figures that are most at risk in a short period of time are those in which the cognitive intervention is minimal. Telephone salesmen (99.0%) are among the professions destined to disappear in the top three positions, followed by typists (98.5%) and secretaries (97.6%). The first place result does not surprise us. Suffice it to say that technology is already sufficiently sophisticated and capable of assisting customers by telephone. With the help of algorithms and data available online, it won't be difficult for bots to try to sell us goods and products over the phone. At the bottom of the rankings and therefore among the jobs destined to resist are those that require analytical skills, empathy, and human relationships. Thus, among the professions most likely to resist automation are educators (0.4%). The awareness is that many jobs will disappear, but also that many others will be born. We can certainly say that the world of the future will appear as a world of an experience economy and an advanced service economy.

## 6.3   ROLE OF UNIVERSITIES IN TRAINING

With a view to digitization, many universities and higher education institutions in the world display their offer in *e-learning* through the so-called MOOCs (Massive

Open Online Courses), i.e. learning platforms that freely offer courses belonging to the official training offer of universities without request the payment of any fee and with the quality level typical of a university. These platforms substantially extend the traditional virtual learning environments (*Virtual Learning Environment* – VLE, also called *Learning Management System* – LMS) and present themselves as large repertoires of the official training offer of a university in *e-learning*, to which anyone can access freely except then having to be accredited as an actual student, paying the relative fees if you wish to obtain the qualification. These systems represent a great vehicle for the dissemination of "certified" knowledge by these institutions and also a very strong marketing element of the "brand" of the best universities in the world. The first MOOC experience, it is worth remembering, is a very famous course in artificial intelligence provided by the University of Stanford.

Training courses are also evolving from the point of view of the provision of content to learners, involving more and more digitization processes. In this context, the main world Business Schools are quickly adapted to the change underway, also from the point of view of connection with students, allowing remote use through an integrated mobile website and responsive, used as a gateway to learning platforms, with the aim of maximizing the usability of the learner experience. In addition, the production of lectures and videos for micro-learning has started, and teaching methods based on new digital technologies have been adopted. More generally, in recent years and with the advent of digitalization, awareness has grown of the fact that it is necessary to continuously review and update skills, which represent the fundamental asset of society in order to meet the needs of the new generations as well. Skills and training will be essential in a job market destined to change more and more due to the digital age. Only through dialogue with the main stakeholders will it be possible to offer cutting-edge tools to promote innovation and the creation of value. Thus, adaptability and innovation are two key factors in guiding the success of any company, and the COVID-19 pandemic has undoubtedly accelerated the innovation and digitization process, representing an opportunity for the conversion of skills and abilities of human capital. Just two years ago, the **World Economic Forum** estimated that, by 2022, more than 54% of employees would be required to undergo significant re-skilling and up-skilling. This time horizon is now decidedly reduced: a real acceleration of the training offer is underway. Today, in fact, thanks to the enormous amount of data available to organizations, it becomes essential to start from big data to address the complexity of the problems that characterize our companies as well as provide a scientific and solid relationship in decision support. We have therefore moved from a disciplinary-based approach to a problem-solving-based one. The use of advanced technologies will make it possible to create hyper-personalized learning paths with the possibility of combining on-site teaching sessions with distance learning moments to stimulate the development of *technological skills*, *creativity*, and *emotional intelligence*. In fact, according to a report from Dell Technologies, 85% of the jobs Generation Z and Alpha will have in 2030 have not yet been invented; it is therefore essential to provide our students with the tools they need to carry out the professions of the future. In this context, Prof. Boccardelli concludes that digital tools make it possible to develop content for remote learning (e.g. provision of blended training courses), facilitating constant learning (so-called continuous learning) which must represent a crucial objective for the training schools.

## NOTES

1 Randstad, Industry 4.0 mismatching and reskilling: https://www.randstad.com/workforce-insights/future-work/industry-40-mismatching-reskilling/
2 The hybrid job economy – How New Skills Are Rewriting the DNA of the Job Market, Burning Glass Technologies, 2019.
3 BBC News Services: Will a robot take your job?: https://www.bbc.com/news/Technology-34066941

# 7 Perspectives on the Future of Digitalization

## 7.1 DIGITIZATION AND INNOVATIVE BUSINESS MODELS

Only in recent years have companies begun to understand the importance of creating environments conducive to success in working life. The idea of work-life integration is gaining more and more ground and it is in this scenario that some new ways of working such as smart working are becoming essential allies for the robust growth of any economic activity. The world events of 2020 in their exceptionality proved disruptive due to the sudden way in which they put before the eyes of entrepreneurs how much digital transformation has been the real, and often, only salvation for companies to cope with the lockdown and to promote new business models.

The COVID-19 emergency has generated a deep rift between the pre-epidemic business model and the one that is gradually being defined and configured in recent months. Just think of the new communication logics that are being established with increasing determination within the same company and that completely bypass the idea of the physical presence of people to make joint decisions or to define even long-term strategies. It is to be hoped that this new working model will be rooted and stable in the DNA of companies in order to build the organizations of the future. We are convinced that "the recent changes triggered by the pandemic have exposed the old continent to an unprecedented awareness and globalization itself has suffered a deadly blow, just think of the increasingly frequent reshoring". The challenges of the future must be based on new ideas on new **models of sustainability**, focusing on the 3 Ps: **profit, planet, and people.** But to achieve these three objectives, a change of mentality is needed, starting with the ability to adopt new competitive and growth logics. A very interesting book in this sense is **Blue Ocean Strategy**, written in 2005 by W. Chan Kim and R. Mauborgne, in which the authors tried to define the new frontiers in the division of the market by two or more competitors. The obsessive and tireless pursuit of conquering market shares to the detriment of those possessed by competitors absorbs a large part of a company's energy in seeking profit in the same market (red ocean), while, as the scholars point out, that same energy could be used energy to think of new still unexplored scenarios (blue ocean). The supermanent of the linear economy also passes through these new logics, favored in many respects by the *digital transformation*, which represents an important differentiating opportunity for companies also in terms of the *circular bioeconomy*. The main reasons are two. The first is linked to the fact that digital transformation processes are usually less energy intensive than traditional ones: just think of smart working (which allows thousands of people to avoid taking the car and therefore consuming fuel to go to work), videoconferencing, or even all the APPs for purchasing products on our mobile phones. The second is

DOI: 10.1201/b22968-7

instead linked to the need to treat the processes of "consumption of resources" with the same logic as the production processes where there is a very strong focus on continuously reducing waste (because they represent important costs). The processes aimed at reducing waste of resources, at this point in a global way, can be accelerated if we use support tools such as those provided by digital transformation.

In light of the previous considerations, we can state that **the evolution towards digitization passes through new business models.**

### 7.1.1 REFLECTIONS ON DIGITAL AND ENVIRONMENTAL INNOVATION AS ECONOMIC DRIVERS FOR A SUSTAINABLE AND RESILIENT SOCIETY

Our society is facing increasingly complex ecological and economic challenges that require a new perspective and collective action to ensure a sustainable and resilient future. In this context, both digital and environmental innovation emerge as a powerful combination capable of driving change towards a greener economy and promoting economic growth.

A tangible example of the impact of digital innovation on the environment is the transition to renewable energy. Through digital solutions, it is possible to monitor and optimize energy efficiency, integrate renewable sources into the power grid, and promote distributed generation. The use of advanced algorithms enables demand prediction, energy distribution optimization, and intelligent resource management. This not only reduces the environmental impact resulting from fossil fuel use but also creates new job opportunities and stimulates economic growth.

On the other hand, environmental innovation involves the development of sustainable solutions to address environmental challenges such as climate change, resource depletion, and pollution. This entails the adoption of clean technologies, the promotion of eco-efficient production processes, and the introduction of sustainable consumption patterns. Environmental innovation requires a long-term vision and collaboration between businesses, institutions, and civil society to develop innovative solutions and promote a culture of environmental respect.

The interaction between digital and environmental innovation can generate positive synergy. Digital technologies provide the tools to collect, analyze, and leverage environmental data, enabling more efficient management of natural resources. At the same time, the focus on environmental sustainability drives digital innovation, as it requires the development of cutting-edge technological solutions to address environmental issues.

Thus, digital and environmental innovation are two fundamental economic drivers for a sustainable and resilient society. Through the integration of digital and sustainable solutions, it will be possible to tackle environmental challenges, promote economic growth, and improve people's quality of life.

The **main challenges** of digital and economic transition are as follows:

- Access and the digital divide: access to digital technologies and internet connectivity is not evenly distributed worldwide. This creates a digital divide between regions, communities, and socio-economic groups, which can hinder digital and economic transition. It is essential to address this challenge through investments in digital infrastructure, education, and accessibility to ensure equitable participation in this transition.

- Cybersecurity: with increased digitization, cybersecurity becomes an increasingly critical challenge. Protecting personal data, preventing cyber threats, and ensuring privacy are fundamental to instilling trust in the digital and economic transition. Robust security policies, data protection measures, and promoting a culture of cybersecurity at all levels are necessary.
- Reskilling: digital and economic transition may involve restructuring of sectors and required skills. Some traditional jobs may become obsolete, while new digital and technological skills become increasingly important. Workforce reskilling and training need to be promoted to ensure people can adapt to the new demands of the job market.
- Social and ethical impact: digital and economic transition can profoundly influence social and economic dynamics. It is important to consider the impact of these transformations on workers, economic inequality, privacy, and concentration of power in the hands of a few actors. Adopting ethical and socially responsible policies and regulations is essential to mitigate negative effects and promote an inclusive and sustainable approach.

Addressing these challenges requires a collaborative approach among governments, businesses, institutions, and civil society. Developing appropriate policies and regulations, investing in education and training, promoting responsible innovation, and ensuring equitable participation are important steps. Only through an integrated and sustainable approach can we successfully address the challenges and seize the opportunities offered by digital and economic transition.

## 7.2 CONCEPTION OF TIME IN THE DIGITAL AGE AND NEW BUSINESS MODELS

If we assume that time and space are conventions established by mankind and conceived, it is easy to understand how digitization is influencing and will continue to influence the way we perceive it. The compression of distances, the irrelevance of geography, *always on* as an existential condition, the conquered simultaneity of processes that not even ten years ago would have required a consistent "jetlag" (the globe is now worth the distance of a tweet), everything leads us to believe that the bits have consumed the real. Intellectuals of caliber such as Manuel Castells, Douglas Rushkoff, and Paul Virilio explain to us that we live in the era of "time without time", of the "eternal now", and of the "tyranny of real time". The same upheaval concerns space. Let's think, for example, of a supermarket. The digital reinvention of the space is evident here: from warehouse management to payment at the tills, every action is computerized to such an extent that if the connection or management software were to fail, that place would stop being a supermarket and become a little more of a warehouse. The code, the connections, and the network spatialize that place to the point of making it what it is. When we say that the code redefines the ontology of space, we mean nothing more than this. Digital immateriality reinvents places and it couldn't be otherwise. From this perspective, space is a constantly evolving creation and the software code is the key element that continuously produces space and its meaning. Scholars of digital philosophy call the generative act of space triggered by software codes the term "transduction". We are facing, indeed, transduced spaces because we live in contexts whose conditions of existence depend on codes and scriptures of reality which transform into a modular becoming what we have always imagined in Newtonian terms as an absolute datum or in Kantian terms as a gnoseological category. The computerization of artifacts therefore generates a renewed temporality and spatiality.

### 7.2.1 A Point of View on Digitization and Innovation: The Circle K Business Model

Standard Cognition, an American startup, was chosen by the Canadian multinational Alimentation Couche-Tarde to develop a pilot project involving the creation of completely autonomous convenience stores. The project is actually based on an idea developed by an Italian start-up, Checkout Technologies, founded in 2017 by Pandian and Jegor Levkovskiy, who developed a technology to eliminate waiting time at supermarket checkouts. Checkout Technologies was then acquired by Standard Cognition for the implementation of the technology in its convenience stores.

The technologies that are used are many and range from artificial intelligence to facial recognition, gesture recognition, and machine learning. The integration of these technologies allows supermarket customers to take the products, pay, and leave the store without going through the checkouts (Figure 7.1).

**FIGURE 7.1**   Circle K, the first fully autonomous convenience store.

In practice, customers autonomously and touchlessly pay via a smartphone app without using cash or credit/debit cards. Customers will be able to walk into the store, grab what they want, and walk out without having to scan anything or stand in line to pay. The novelty is that no sensors will be used but cameras mounted on the ceiling will be used. Furthermore, through artificial intelligence algorithms, each buyer will be accurately associated with the items he collects without using biometric data.

### 7.2.2 A POINT OF VIEW ON DIGITIZATION AND THE CONCEPT OF TIME: THE DECISION LENS

We explored the topic of Digitization and the Concept of Time with **John Saaty**, co-founder and CEO of Decision Lens, a software company based in Washington (USA), whose mission is to provide consultancy and solutions for effective planning, prioritization, and optimization of resources.

---

**Co-founder and CEO of Decision Lens**

In 2005, Decision Lens founders John and Dan Saaty launched their father Dr. Thomas Saaty's world-renowned decision framework as powerful software that would revolutionize the way the public sector prioritizes, plans, and funds.

Decision Lens is integrated planning software that modernizes how the government prioritizes, plans, and funds. Leveraging our unique expertise in decision science, agencies achieve a sustained operational advantage through superior long-term planning, continuous medium-term prioritization, and short-term funding execution.

---

Among the many benefits that digitization brings to companies, one of the most overlooked is assumptions about work processes and how they are affected by time. It used to be that the work processes in companies and the cycles for executing those processes were dictated by the amount of time that it took to bring people together, do the analysis, collaborate, evaluate the way forward, commit to a particular agenda, and then execute. All of this has been upended by digitization.

Digitization can compress time itself by enabling nearly instantaneous sharing and analysis of information, with the use of machine learning and algorithms to immediately offer pathways forward, or it can make time nearly irrelevant through things like contemporaneous collaboration amongst parties where not previously possible. The result that we are seeing play out in front of us today is that business cycles have sped up and, in some cases, disappeared entirely.

Take the planning cycle as an example. In the latter half of the 20th century, it used to be accepted that companies would develop an annual plan and then execute the plan. It was a laborious process of meetings, engagement, pulling data together from manually reported sources, arranging the data and analyzing it, meeting more to evaluate the results, and then coming to a conclusion and reporting recommendations. The process was so involved and top-heavy that once you set the planned path forward, you executed that plan almost regardless of what new information came to light over the course of the year. Employees put their heads down and drove towards the stated goal, and then lifted their heads to assess the environment, ecosystem, competitors, and technical progress at the end of the year in order to begin the new cycle.

The first wave of digitization in the 1980s and 1990s sped up this cycle by leveraging servers that enabled the aggregation of data through databases, and desktop PCs provided the capability to analyze that data in spreadsheets or statistical programs to draw conclusions and present to leadership. Companies found that they could interject a second "semi-annual" planning cycle in the year and redirect resources where possible.

As the Internet developed and most systems moved to the cloud, and as analytic capabilities advanced with increased computing power, the availability of the data and analysis increased tremendously in velocity. Additionally, it became much simpler to collaborate and move forward through virtual engagement. Companies realized that they were no longer shackled to a semi-annual planning process, but instead moved it to monthly, then semi-monthly, then to weekly "respins" of the plan, and now many companies are realizing the benefits of what is being called "continuous planning" where a group of employees is dedicated to driving the planning, analysis, and resource allocation and re-allocation on a constant basis.

The occurrence of the pandemic and COVID-19 in 2020 caused this cycle to speed up even more. Because people were forced out of the office to work remotely, they increasingly relied on cloud-based systems to run their organizations. The leftover relics of manually-based processes or "status quo" – where they continued to implement a particular process "because that's how it's always been done" went out the window. Like the concept of sunk cost, it made no sense to try to perpetuate business processes that were designed for a non-digitized world when it wasn't even possible anymore to enable those processes. While the early adopters had long since taken advantage of digitization, the early majority and late majority companies in the adoption cycle were forced to jump across the chasm and change their thinking and processes nearly instantaneously to a digitized world.

The benefits are many – organizations are now able to nearly instantly identify and take advantage of new trends, react immediately to competitive moves, get products out rapidly through faster, digitized supply and distribution chains, and more. In many ways, this has leveled the playing field amongst firms and price competition has benefited consumers enormously. Many staple products cost less today on an inflation-adjusted basis than they did 30 years ago. And these benefits keep coming. This is because "time" has been dramatically reduced as a factor in business cycles through the power of digitization.

## 7.3  BUSINESS OPPORTUNITIES OF THE DIGITAL BIOECONOMY

In our opinion, digital transformation and sustainable development are two Italian and European objectives that must be pursued together. In our opinion, the problems related to the distribution of resources, in addition to those related to over-consumption, should make us reflect on the fact that we tend to consume more than we actually need. The global spread of COVID-19 has forced many countries to slow down their activities, but there have also been those, like many countries affected by the pandemic, who have had to stop. Precisely in these complex moments, we have witnessed a reduction in road traffic due to an almost total use of smart working and teleworking. At the same time, the lockdown, as opposed to the concept of the consumer economy, has taught us that we need much less to live. From now on, each of us should reflect in terms of the environmental impact of our business. We should learn to ask ourselves *what can I improve, to make a difference globally?* Only in this way will we fully understand the potential of digital, which allows people to bring about small changes at the individual company level, but capable of generating a real social impact. Only in this way will we be aware of the fact that everyone's contribution is fundamental. Changing our habits will lead economies to focus more and more on **sustainability** and the **circular economy**. A significant litmus test of the ongoing process of change is what is happening to oil. The raw material par excellence of the 20th century, protagonist of the first two industrial revolutions and symbol of the power of states and multinationals, today no longer seems indispensable. There is more than what is needed, and although it seems incredible, its price in the period of the "global closure" has been quoted as negative. Nobody wanted it and people were willing to pay just to store it some-where; 30 or 40 years ago, nobody would have ever thought of it. It is therefore necessary and desirable for companies to accelerate internal change while at the same time acting as a driving force to give a boost towards digitization and sustainability. Even companies linked to "traditional heavy industry" processes are strongly committed to this direction.

## 7.4  SMART WORKING AND THE WORLD OF WORK

By now, it is known that the "lockdown" has highlighted that most companies did not have structured initiatives – others still had not even foreseen the activation – to work in "smart working". At the stroke of the lockdown, an extreme measure against an invisible enemy, technology became the anchor to grasp in a socio-economic scenario which, week after week, sailed on sight, unable to reach a safe and forced port to adapt to a life made up of mobile horizons. According to research conducted by the International Labor Organization (ILO) in 2020, during the COVID-19 pandemic, up to 30% of workers globally engaged in some form of smart working, such as working from home or telework. It was an epochal change, which undermined the still too often rigid foundations of corporate cultures in which control of the work of one's own resources maintains primacy over their involvement in the corporate project. We are facing an opportunity for companies and resources to build collaborations based on common intentions, values, and visions, which however brings unresolved and burning questions to the table, determined by the redefinition of spaces, times, and ways of working.

Indeed, what weakens the potential of smart working is a real aporia of the digital age: the ease with which, as individuals, we are led to implement different and varied variations of technologies in our daily lives, as opposed to the slowness and lack of inclination to transfer them within the professional, entrepreneurial, and productive systems. An aporia that can certainly be explained by the existence of bureaucratic and procedural brakes within organizations, but which finds its deepest roots, perhaps, in the lack, particularly in our territory, of a corporate and en-trepreneurial culture oriented towards change and innovation, perceived as a threat to the *status quo* rather than an opportunity for growth. Of course, "agile" work is not the panacea for all ills, but it is an investment of resources and time necessary for the reorganization of all business processes. It is a challenge that could improve the quality of life of thousands of workers and the productivity of businesses. We have the **opportunity to experiment with** new balances, designing "blended" ways of working, which alternate office life and remote activities. Because if it is true that studies show that, in terms of productivity, smart working represents an added value, it is equally true that the physical isolation of the worker risks resulting in an isolation of ideas and, consequently, in an impoverishment of "creative entropy that arises only from the experience of collaboration". Smart working impacts work and corporate organization, but not only. In recent months in the United States, in San Francisco, thousands of employees of technology companies struggling with remote working to the bitter end have decided to leave their rented homes to move else-where, where the rental costs are much lower. The lockdown has opened up new scenarios: living in less urbanized or rural areas by reconciling family, profession, and free time. Meanwhile, companies are wondering about the reduction in salaries. Those who live in Silicon Valley usually, precisely because of the high cost of living, receive a higher salary than a peer who works in other areas of the United States, and some workers have accepted a reduction in salary in order to live in areas less expensive, where the quality of life is however higher. Probably in the not-too-distant future, fewer employees will physically go to work "in the office", and the very concept of the office will expand, taking into account co-working and smart working spaces. To paraphrase **Zigmunt Bauman's** famous metaphor, work is becoming liquid at all levels and in every place, even if there are exceptions. There are countries that do not accept the **modern world**. One of them is **Belarus** where **typewriters** are in use in a large percentage of businesses and government offices.

## 7.5   THE METAVERSE: A NEW FRONTIER OF DIGITIZATION

The term "metaverse" refers to a virtual shared space where users can interact, communicate, and engage with one another in immersive and digital environments. The term was coined by Neal Stephenson in the cyberpunk novel Snow Crash (1992) to indicate a three-dimensional space within which natural persons can move, share, and interact through custom.

The term, a union of *meta* (inside) and *toward* (abbreviation of "universe"), indicates a universe parallel to the real world.

The birth and evolution of the metaverse represent one of the most fascinating and promising frontiers of the digital world. The concept of the metaverse has deep roots in the history of virtual reality and interactive virtual environments, but it is in recent years that it has gained greater attention and relevance. We then explore the history and evolution of the metaverse, as well as the opportunities and challenges it presents for the future.

The birth of the concept of the metaverse can be traced back to the 1980s, when visionaries like **Jaron Lanier** began imagining virtual worlds where people could interact and communicate in an immersive way. However, it has been the recent convergence of technologies such as virtual reality, augmented reality, artificial intelligence, and network connectivity that has accelerated the evolution of the metaverse.

In recent years, we have seen growing interest in and increased investment in the emerging technologies needed to create the metaverse. Companies like Facebook, Microsoft, Google, and others have devoted significant resources to developing platforms and applications that bring the metaverse closer to reality.

One of the most exciting aspects of the metaverse is the ability to create immersive social and interactive experiences. In the metaverse, people can interact with each other, communicate, collaborate, and share experiences in entirely new ways. This opens the door to extraordinary opportunities in education, entertainment, collaborative work, and commerce.

In education, the metaverse can provide immersive and interactive learning experiences, allowing students to explore virtual places, interact with 3D objects, and participate in realistic simulations. This could revolutionize the way we learn, making education more accessible, engaging, and personalized.

In entertainment, the metaverse can transform the experience of playing video games, offering expansive and immersive virtual worlds in which players can explore, socialize, and participate in shared adventures. This opens up new possibilities for competitive gaming, esports, and virtual entertainment experiences.

However, while the metaverse offers tremendous opportunities, there are also significant challenges to its full development and widespread adoption. Figure 7.2 summarizes the main applications of Metaverse.

One of the main challenges is to create an open and interoperable metaverse. Currently, there are several platforms and virtual worlds, each with its own rules and standards. To realize the full potential of the metaverse, it will be necessary to work towards common standards that allow for interaction between different platforms and a cohesive experience for users.

Security and privacy are other major challenges. In the metaverse, people create and share sensitive personal data, interact with other users, and participate in digital transactions. Therefore, it is essential to ensure that user information is protected and that there are robust mechanisms in place to manage security and privacy in the context of the metaverse.

Finally, the metaverse raises questions of inclusiveness and accessibility. To ensure that the metaverse is an inclusive environment, different abilities and

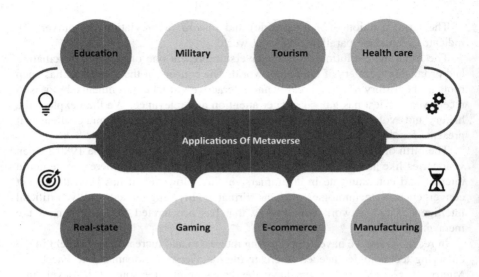

**FIGURE 7.2**   Main applications of Metaverse.

economic barriers need to be taken into account. This will require access to appropriate devices and internet connections, as well as adopting accessible design standards to enable all people to participate in and benefit from the metaverse.

In conclusion, the metaverse represents a new frontier that offers extraordinary opportunities in the fields of education, entertainment, work, and commerce. However, to fully exploit the potential of the metaverse, significant challenges such as interoperability, security, privacy, and inclusiveness will need to be addressed. If these challenges are tackled collaboratively and responsibly, the metaverse could open up new horizons for human interaction and the creation of digital experiences that enrich our lives in ways we have never imagined yet.

## 7.6   GLOBAL CHALLENGES FOR OUR FUTURE

As we witness the latest gimmick from the eCommerce giant that has just launched a new service (for now only in the States) called **Amazon Key** with the intention of being able to make deliveries even when the user is away from home, we wonder what will be the challenges of the near future and what will be the opportunities. Probably, our future will be guided by a few key words: **genomics**, **biology**, **robotics,** and **cybernetics**. All powered by Big Data and artificial intelligence. The new prospects for development opportunities in the quality of life and in the world economy will be based on specific and sectoral skills that will give the possibility of knowing what to achieve and how to achieve it. The strategies of each country should aim to **compensate** for their **weaknesses** rather than consolidating skills. A commitment to fostering an innovative entrepreneurial culture based on openness to become a global investment hub should be a key requirement for any successful company. In a seminar focused on innovative technologies, **Gideon Lichfield**, Editor-in-Chief of MIT Technology Review, told us how, from his point of view, the health emergency has acted as an accelerator not only of technology, but above all of culture. For example, **Microsoft** is working on a new platform aimed at changing the customer experience in meetings, providing solutions capable of simulating a habitat in which participants have the sensation of being present all together in the same space of the "infosphere". MEEPL Company from Zürich (Switzerland) has developed a technology capable of reproducing very accurately virtual bodies in 3D and models of clothing capable of giving the consumer a new "digital experience". The app in question allows you to simulate a real "dress fitting", entering your height, weight, and other characteristics in the generated avatar (https://www.meepl.com/meepl).

The health emergency, underlines Gideon Lichfield, is "forcing" us to have more faith in technology. It is known that innovation suffers from inertia, as we have already had the opportunity to underline; furthermore, it is human nature to go in search of a solution that allows us to remain in a state of comfort.

Alongside the acceleration of trends that can no longer be postponed, the emergency has also reinvigorated us with crucial questions about futuristic scenarios. **Lichfield,** for example, is convinced that success in the near future will be achieved by combining new technologies or artificial intelligence with the traditional methods of decision making. More specifically, it asks: *will there be a physical body for artificial intelligence?*

According to **Peter Diamandis,** we are close to the same revolution that occurred more than 3 million years ago, when we passed from the eukaryotic cell to the morula, or rather, an aggregate of 12 cells, the embryo of the human body. Today, after several million years, we are about to witness an epochal technological innovation that until a few years ago would have seemed like science fiction: **rewriting the DNA of which the genome of all living organisms is made up**. Any type of plant or animal cell, including the human one, can be genetically modified, and the correction can take place even for a single and minimal error and anywhere in the genome. Among the pioneers of these studies is the US scientist **Jennifer Doudna** of the University of Berkeley, who revolutionized genetic engineering by winning the Nobel Prize for Chemistry in 2020. Thanks to artificial intelligence, it has been demonstrated that the modification inserted by the "kit of repair" of the cell is predictable. In other words, AI algorithms are able to predict how the cell will repair the DNA molecule after it has undergone editing. A potential that leads us to think that we are close to a **new singularity**.

## 7.7 A QUESTION OF "ETHICS": TOWARDS ALGORITHMS

Technologies such as Big Data and Artificial Intelligence have brought *privacy* and *ethical issues* related to the application of algorithms back to the world's attention. There is more and more talk of **algorithmics**. The existing regulatory gaps determine the emergence of phenomena of abuse of power and lack of transparency. Combining ethics and technology for new technologies that always place man at the center and are at the service of authentic development could represent a perspective to be pursued for the protection of all citizens. But new criteria, categories, and languages are needed. When we talk about new technologies, we are basically talking about algorithms. In the international debate, reference is often made to *ethics by design* to describe the alignment with ethical principles of the whole process of conception, development, and implementation of an artificial intelligence system. It was **Donn Parker**, in his *Rules of ethics in information processing*, who outlined the idea of data processing that could be designed according to ethical principles. In the context of the debate on the ethical and legal implications of the behavior of robots, there is, among others, the issue of the processing of **personal data**. Think of employment in the personal care sector, where it cannot disregard the processing of personal data. Such a scenario poses an issue related to the introduction of so-called *apps on the market*. One aspect that was at the center of attention in the news related to the COVID-19 health emergency was that of **tracing the movements** of affected people and their contacts via apps. Precisely for this reason, even when we talk about apps that are born with "social" purposes and for the protection of the community, such as apps for COVID-19, the installation of the application is not mandatory, but takes place on a voluntary basis. In any case, the issue is delicate. There is still a lot to do.

# Conclusions

## THE DIGITAL AGE BETWEEN TECHNOLOGY, SKILLS, AND OPPORTUNITIES

The new information technologies are not limited to changing our habits and our daily lives. Digital technologies are changing the world. It may sound strange, but it is not such a new idea. It began to flash already in the mid-20th century, when the information revolution triggered by **Alan Turing** and his audacious and shocking idea of being able to manufacture *thinking machines*, led scientists and philosophers to reconsider the very ontology of things. **Sir John Archibald Wheeler**, for example, a giant of 20th-century physics, pioneer of quantum gravity, black hole theorist, and privileged interlocutor of Albert Einstein, Niels Bohr, and Richard Feynman, in the last part of his career, became convinced of the possibility of being able to bring the entire physical world back to the bit. *Everything is information. The more I reflect on the mystery of quanta and our unique ability to understand the world we live in, the more I become convinced that logic and information may play a vital role in the foundation of physical theory.* The bit, the information, becomes the very principle of reality, an idea that has gradually crept into the minds of all those who would go on to write the history of information technology in various ways. Already towards the end of the '60s, the inventor of the first programmable computer, **Konrad Zuse,** wondered, for example, whether nature was analogical or digital (opting for the second hypothesis), a question resolved definitively in the famous conference that, in May 1981, saw the birth of the *Physics of Computation* at the Massachusetts Institute of Technology. In the great hall of the legendary Endicott House, the MIT convention center, in the presence of some of the most brilliant minds of the 20th century, Freeman Dyson, Norman Packard, Carl Adam Petri, Hans Moravec, and Gregory Chaitin, they all agree with Zuse and Wheeler: the universe is computation.

The digital revolution is so profound that it invests the very essence of reality. There is little wonder. Vico, Galileo, and then Kant with his *Copernican revolution* had already clarified that the reality of things is conditioned by the way we know them. In other words, the world as we know it is the synthesis of our ability to negotiate with it through codes and languages, codes and languages that, if they were mathematicians from Descartes and Galileo, starting from Turing, began to be predominantly computerized. *The world is a flow of signs,* wrote Peter Weibel, director of the ZKM, the Center for New Art and Technology in Karlsruhe, and one of the most acute observers of the contemporary world. In other words, the world is an infosphere, a concept reworked and made famous in his own way in recent years by Luciano Floridi, professor of Ethics and Philosophy of Information at the University of Oxford. *We are witnessing an epochal and unprecedented migration of humanity from Newtonian physical space to the new environment of the*

DOI: 10.1201/b22968-8

*infosphere*, writes Floridi. And yet, more than 70 years after Turing's article, our conceptions of the ultimate nature of reality are still *Newtonian,* in the sense that we still think of cars, buildings, furniture, clothes, and objects of all sorts as dumb things, unable to interact, respond, learn, or memorize. However, this world is gradually dissolving; it is hybridizing with environments animated by widespread information processes, distributed and operating at any time and in any place. You live *onlife* and no longer just *onsite*. Digital technologies slowly change our vision of mechanical technologies. "At the end of this passage – Floridi prefigures – the infosphere will no longer be conceived as a way of referring to the information space, but as a synonym of reality itself".

Here, today the game is played here, in the **infosphere**: either we are able to govern it or we end up suffering it. We experienced it firsthand during the lockdown. When schools and offices closed, only those who had the privilege of living in the infosphere were able to continue taking courses, taking exams, interacting with colleagues and customers, and selling products. For those who are inside the digital sphere, the presence is detached from the locality, becomes lighter, and can travel on the fiber without anyone necessarily moving from home or office. In the analogue world, however, locality and presence are inseparable and those who were forced out of digital during the months of forced closure were literally kicked out. He stayed out of training, work, and even social life (video calls on Houseparty are not the best but better than nothing!). In other words, outside the infosphere, one is not simply outside the digital world; one is outside the world tout court. That digital innovation is distributed in a capillary manner is therefore a question of civilization, even before development opportunities. It is not enough to build the large (and indispensable) backbones. Whether we are talking about mobility or infrastructure, what is still missing is the coverage of those last few hundred meters that bring technology and the benefits that derive from it into people's homes and workplaces. Only by bridging this gap will it be possible to ensure not only business continuity, but also the maintenance and growth of productivity in a context where the line between work and private life is increasingly blurring. As **Kevin Kelly** had predicted in *The Inevitable. The technological trends that will condition our future,* everything has become fluid: presence, for example, has freed itself from "locality", just as distance has freed itself from time. The increasingly evanescent boundaries of the workplace as well as the mixing of times dedicated to productivity and free time, considered absolutely incompatible only a few years ago, are nothing more than a consequence of this, indeed, "inevitable" trend.

The technologies enabling the Digital Revolution, such as 5G, beyond the glossy images imprinted in the collective imagination of surgeons who "virtually" operate on patients hundreds of kilometers away, will be able to reduce latency not only in our connections but above all in our lives, redesigning our way of conceiving time and space. It is a change that will have a major impact on companies, on the way of producing wealth and well-being, and, naturally, on local areas. Now we all know the reality of smart working, which is not the simple relocation of home work, but a restructuring of functions within an overall reorganization of production processes. Well, at the heart of this reorganization is the possibility of finally thinking in terms of objectives. The company is no longer interested in *buying* its collaborator's time;

rather, it relies on their abilities. All of this could be a wealth multiplier and it is no coincidence that companies are reorganizing themselves in this direction. It is a question of knowing how to rearticulate the *utilization flows* of capital (premises, tools, plants) and of workers, and this, together with the individual companies, will inevitably lead to a transformation of the production chains as well. Thanks to the digital liberation of the presence from the locality, many families, for example, will be able to imagine returning to repopulate the provinces. For the first time in decades, we will be able to witness a movement in contrast with urbanization and gentrification, to which the redevelopment and regeneration concept. Other changes will be part of a broader drive towards more sustainable lifestyles, not only from an environmental point of view, but also from a social and economic point of view.

What we are experiencing is an **epochal moment**. Just as Galileo's mathematics was the language of physics with which Newton laid the premise of the industrial revolutions, Turing's information technology is the language with which we are now called to interpret and carry out the digital transition. "Thanks to Turing – writes Floridi again – Bacon and Galileo's project to grasp and manipulate the alphabet of the universe has begun to find realization in the computational and informational revolution, which is so deeply influencing our knowledge of reality and the way we represent it and conceive ourselves within it". In the 1950s, human beings walked inside computers because computers were as big as rooms. Today, we walk inside computers again, but in a completely renewed sense. Distributed artificial intelligence, virtual reality, industrial and service robotics, cobots, cyber-physical environments – all this is transforming the way we work, do business, study, and innovate in an increasingly inextricable interaction between natural and artificial cognitive agents. The "thinking machines" are among us, and, together with us, they are tracing the boundaries of the new world.

# Bibliography

Alec Ross - *"The Industries of the Future"* (2016, Simon & Schuster).

Alex Osterwalder and Yves Pigneur - *"Value Proposition Design: How to Create Products and Services Customers Want"* (2014, Wiley).

Amy Webb - *"The Signals Are Talking: Why Today's Fringe Is Tomorrow's Mainstream"* (2016, PublicAffairs).

Bernard Marr - *"Artificial Intelligence in Practice: How 50 Successful Companies Used AI and Machine Learning to Solve Problems"* (2019, Wiley).

Brian Christian - *"The Alignment Problem: Machine Learning and Human Values"* (2020, W. W. Norton & Company).

Bruce Schneier - *"Click Here to Kill Everybody: Security and Survival in a Hyper-connected World"* (2018, W. W. Norton & Company).

Cal Newport - *"Deep Work: Rules for Focused Success in a Distracted World"* (2016, Grand Central Publishing).

Cal Newport - *"Digital Minimalism: Choosing a Focused Life in a Noisy World"* (2019, Portfolio).

Charlene Li - *"The Disruption Mindset: Why Some Organizations Transform While Others Fail"* (2019, Ideapress Publishing).

Christian Madsbjerg - *"Sensemaking: The Power of the Humanities in the Age of the Algorithm"* (2017, Hachette Books).

David Epstein - *"Range: Why Generalists Triumph in a Specialized World"* (2019, Riverhead Books).

DIGITAL - *"Global Digital Overview Essential Insights into How People around the World Use the Internet, Mobile Devices, Social Media, and Ecommerce"* (2023, We are social hootsuite).

Eric Ries - *"The Startup Way: How Modern Companies Use Entrepreneurial Management to Transform Culture and Drive Long-Term Growth"* (2017, Currency).

Erik Brynjolfsson and Andrew McAfee - *"Machine, Platform, Crowd: Harnessing Our Digital Future"* (2017, W. W. Norton & Company).

European Commission - *"Directorate-General for Research and Innovation 100 Radical Innovation Breakthroughs for the future"* (2019, Luxembourg: Publications) ISBN 978-92-79-99139-4.

Gary Marcus and Ernest Davis - *"Rebooting AI: Building Artificial Intelligence We Can Trust"* (2019, Vintage).

Gary Vaynerchuk - *"Crushing It!: How Great Entrepreneurs Build Their Business and Influence-and How You Can, Too"* (2018, Harper Business).

Geoffrey G. Parker, Marshall W. Van Alstyne, and Sangeet Paul Choudary - *"Platform Revolution: How Networked Markets Are Transforming the Economy—and How to Make Them Work for You"* (2016, W. W. Norton & Company).

Greg Satell - *"Mapping Innovation: A Playbook for Navigating a Disruptive Age"* (2017, McGraw-Hill Education).

Kai-Fu Lee - *"AI Superpowers: China, Silicon Valley, and the New World Order"* (2018, Houghton Mifflin Harcourt).

Karl Popper - *"In Search of a Better World: Lectures and Essays from Thirty Years"* (1995, Routledge Taylor&Francis).

Markus Christen, Brent Mittelstadt, and Luciano Floridi - *"AI Ethics: The Birth and Rise of Autonomous Agents"* (2020, Oxford University Press).

Martin Ford - *"Architects of Intelligence: The Truth about AI from the People Building It"* edited by (2018, Packt Publishing).

Martin Ford Rise of the Robots - *"Technology and the Threat of a Jobless Future"* (2015, Basic Books).

Martin Lindstrom - *"Small Data: The Tiny Clues That Uncover Huge Trends"* (2016, St. Martin's Press).

Matthew Ball - *"The Metaverse: What It Is, Where to Find It, Who Will Build It, and Fortnite"* (2020, Metaversal).

Melanie Mitchell - *"Artificial Intelligence: A Guide for Thinking Humans"* (2019, Farrar, Straus and Giroux).

Michael Negnevitsky - *"Artificial Intelligence: A Systems Approach"* (2020, CRC Press).

Neal Stephenson - *"The Metaverse: An Infinite Reality"* (2022, William Morrow).

Nick Polson and James Scott - *"AIQ: How People and Machines Are Smarter Together"* (2018, St. Martin's Press).

Nir Eyal - *"Hooked: How to Build Habit-Forming Products"* (2014, Portfolio).

Peter H. Diamandis and Steven Kotler - *"The Future Is Faster Than You Think: How Converging Technologies Are Transforming Business, Industries, and Our Lives"* (2020, Simon & Schuster).

Reid Hoffman, Chris Yeh, and Bill Gates - *"Blitzscaling: The Lightning-Fast Path to Building Massively Valuable Companies"* (2018, Currency).

Ryan Holiday - *"Perennial Seller: The Art of Making and Marketing Work That Lasts"* (2017, Portfolio).

Safi Bahcall - *"Loonshots: How to Nurture the Crazy Ideas That Win Wars, Cure Diseases, and Transform Industries"* (2019, St. Martin's Press).

Safi Bahcall - *"The Innovation Stack: Building an Unbeatable Business One Crazy Idea at a Time"* (2020, St. Martin's Press).

Satya Nadella - *"Hit Refresh: The Quest to Rediscover Microsoft's Soul and Imagine a Better Future for Everyone"* (2017, Harper Business).

Scott D. Anthony - *"Dual Transformation: How to Reposition Today's Business While Creating the Future"* (2017, Harvard Business Review Press).

Scott Hartley - *"The Fuzzy and the Techie: Why the Liberal Arts Will Rule the Digital World"* (2017, Houghton Mifflin Harcourt).

Simon Sinek - *"The Infinite Game"* (2019, Portfolio).

Steven Johnson - *"Farsighted: How We Make the Decisions That Matter the Most"* (2018, Riverhead Books).

Stuart Russell - *"Human Compatible: Artificial Intelligence and the Problem of Control"* (2019, Viking).

Thomas H. Davenport - *"The AI Advantage: How to Put the Artificial Intelligence Revolution to Work"* (2018, MIT Press).

Tim Brown - *"Change by Design: How Design Thinking Transforms Organizations and Inspires Innovation"* (2019, Harper Business).

Tim O'Reilly - *"WTF?: What's the Future and Why It's Up to Us"* (2017, HarperBusiness).

Tom Chatfield, Julian Bleecker, Elizabeth Renieris, and Dean Takahashi - *"The Metaverse: An Essential Guide"* (2023, Basic Book).

Tom Taulli - *"Artificial Intelligence Basics: A Non-Technical Introduction"* (2020, Apress).

Vivek Wadhwa and Alex Salkever - *"The Driver in the Driverless Car: How Our Technology Choices Will Create the Future"* (2017, Berrett-Koehler Publishers).

W. Chan Kim and Renée Mauborgne - *"Blue Ocean Strategy: How to Create Uncontested Market Space and Make the Competition Irrelevant"* (2005, Harvard Business Review Press).

Whitney Johnson - *"Build an A-Team: Play to Their Strengths and Lead Them Up the Learning Curve"* (2018, Harvard Business Review Press).

# Index

Printed in the United States
by Baker & Taylor Publisher Services